SIX BRIDGES *The Legacy of Othmar H. Ammann*

SIX

BRIDGES

THE LEGACY OF OTHMAR H. AMMANN

Darl Rastorfer

Yale University Press

New Haven and London

FRONT-MATTER ILLUSTRATIONS

Photograph of Othmar H. Ammann, 1931;

    courtesy of Margot Ammann-Durrer

George Washington Bridge, view from the pedestrian

    walkway; photo, Jet Lowe, courtesy of the Library

    of Congress

George Washington Bridge, suspension cables with

    roadway below; photo, Dave Frieder

Triborough Lift Bridge, detail of tower; photo, Dave Frieder

Bayonne Bridge, pedestrian walkway; photo, F. S.

    Lincoln, courtesy of the New York Public Library

Designed by Richard Hendel

Set in City and Monotype Garamond type by

B. Williams & Associates

Printed in Hong Kong by C & C Offset Printing

Library of Congress Cataloging-in-Publication Data

Rastorfer, Darl.

    Six bridges : the legacy of Othmar H. Ammann / Darl Rastorfer.

        p.   cm.

    Includes bibliographical references and index.

    ISBN 0-300-08047-6 (alk. paper)

    1. Bridges—New York Metropolitan Area—Design and construction. 2. Bridges, Long-span—New York Metropolitan Area Pictorial works. 3. Ammann, Othmar Hermann, 1879–1965. I. Title.

    TG25.N5R37   2000

    624'.2'097471092—dc21   99-39824

A catalog record for this book is available from the British Library.

The paper in this book meets the guidelines for permanence and durability of the Committee on Production Guidelines for Book Longevity of the Council on Library Resources.

10 9 8 7 6 5 4 3 2 1

To my mother and father with gratitude

Contents

Preface, ix

1 Othmar H. Ammann: The Man, His Legacy, 1

2 The George Washington Bridge, 39

3 The Bayonne Bridge, 77

4 The Triborough Bridge, 93

5 The Bronx–Whitestone Bridge, 115

6 The Throgs Neck Bridge, 125

7 The Verrazano-Narrows Bridge, 135

Appendix: Ammann's Built and Unbuilt Projects, 163

Notes, 177

Glossary, 179

Selected Bibliography, 181

Acknowledgments, 183

Index, 185

Preface

It was my good fortune to meet several individuals who had formerly been associated with Othmar H. Ammann. They argued that Ammann was an unknown American master; I came to agree with them. Like many people, I had been aware of the bridge designer's role in the George Washington and Verrazano-Narrows Bridges, but was surprised when I learned that he had engineered four additional long-span and three short-span bridges in the New York metropolitan region. All are dazzling structures. Knowing that, it wasn't surprising to learn that Ammann had also introduced a number of significant innovations to design theory and practice. His phenomenal accomplishments appeared equal to those of any bridge designer in the twentieth century. I was eager to find out more about the structures and more about the engineer.

Learning about the bridges Ammann designed was relatively easy. He wrote and published exacting descriptions of their planning, design, and construction. By reading his writing and that of others, visiting the bridges, and speaking with the dedicated individuals responsible for their daily operation, all those tons of powerfully aligned concrete and steel became more distinct and familiar.

Getting acquainted with the man behind the bridges was a far less simple undertaking. Ammann had a great deal to say about engineering, but very little to say about himself. When journalists asked about his personal life, he always deferred, steering the topic back to bridges. He was equally taciturn with colleagues and friends. One can admire stoic reserve in an individual, but such people pose difficulties to biographers. Fortunately, soon after I began work on this book, boxes filled with personal letters came into the possession of the engineer's daughter, who, in turn, made them available to me. This correspondence among Ammann, his first wife, and his parents presented whole new facets to his life and personality, and served to transform a historic figure known only by name into a living, breathing character. Just as there is much to admire in Othmar H. Ammann's extraordinary engineering work, there is much to admire in the man. May the reader, like me, be inspired by both.

1.1 Self-portrait by Othmar H. Ammann in 1904, soon after he arrived in the United States. The twenty-five-year-old engineer mused for the camera with no inkling of the extraordinary opportunities that awaited. (Photo courtesy of Margot Ammann-Durrer)

Othmar H. Ammann *The Man, His Legacy*

During a long life of uninterrupted practice, Othmar H. Ammann designed and built a series of long-span bridges that altered the course of engineering history and transformed the face of New York. He also wrote some of the finest technical papers ever written in his field and founded two public departments of civil engineering and two private consulting firms. This highly accomplished man was surprisingly self-effacing. When thrown in the spotlight of fame, his response was to maintain the quiet, ordered lifestyle he had adopted during the lean years of his early career: never stay late at the office; be home in time for dinner; relax and then work some more in the evenings and on weekends; garden, hike, and listen to symphonic music for recreation. His focus rarely strayed from family, home, and the design of the world's longest bridges.

Ammann never intended to be one of those rare civil engineers who gains celebrity status, but his premiere as a designer at forty-four transformed his career. Presented with an extraordinary design challenge, Ammann's genius and formidable knowledge made him equal to the job, as he broke with the past, seized the opportunity, and emerged as a master builder with the design of the George Washington Bridge.

Mere size and proportion are not the outstanding merit of a bridge; a bridge should be handed down to posterity as a truly monumental structure which will cast credit on the aesthetic sense of present generations.
—*Othmar H. Ammann, 1954*

BEGINNINGS

On the morning of May 5, 1904, Ammann disembarked from his first transatlantic voyage and set foot on the island of Manhattan. A recent graduate of one of Europe's most prestigious schools of engineering—the Swiss Federal Polytechnic Institute in Zurich—Ammann had traveled to America with the intention of spending one or two years as an employee of engineers who specialized in the design of steel bridges. A year earlier, several former classmates with similar ambitions had established themselves in New York. They helped Ammann gain his bearings amid the city's frenetic whirl. In a matter of days he was established in a boarding house and working through a list of prospective employers.

Ammann recorded his first impressions of the city in letters to family and friends. Of New Yorkers he observed, "What struck me immediately was the healthy, characterful appearance of the people." As for his initial impressions of the city's thoroughfares, architecture, and ceaseless drive, he reported,

> The street traffic astonished me no less than the harbor traffic. In that respect the European cities are way behind. One does not find many beautiful trees, as in Paris, for example; but instead, the most daring buildings, some up to 30-story-high "skyscrapers," that in their own way are highly interesting and thrilling.
>
> . . . So much has passed my eyes in the past few days that I would like to retreat to a quiet corner in order to digest all these impressions with pleasure. But now I must press forward, neither resting nor pausing, each moment must be used. Somehow it becomes apparent that time is precious, everyone rushes and pushes, everything is in a busy excitement and everyone is working for a goal without caring about the rest of mankind.[1]

Ammann's decision to venture to the United States was influenced by a teacher, Karl Emil Hilgard, who had worked in the United States between 1893 and 1896 as a bridge designer for the Northern Pacific Railroad. His enthusiasm for the American scene was infectious, and Ammann fell under its spell. Hilgard urged him to work there for a period, asserting that a young engineer could be given responsibilities that were available only to "gray beards" in Europe. Upon graduation in 1902, Ammann briefly worked in Switzerland before taking a job with an acclaimed firm in Frankfurt. Restless in spirit and eager to excel, a year later he was ready to act on Hilgard's advice.

It took Ammann less than two weeks to find his first position in New York, where he was hired by the office of Joseph Mayer, a consulting engineer. This widely respected practice, located at 1 Broadway, represented a splendid opportunity for the twenty-four-year-old engineer, as it specialized in the design of long-span steel bridges. As Ammann explained in a letter to his girlfriend in Zurich, Lilly Selma Wehrli, "a consulting engineer is an independent professional who provides expert consultation for difficult projects undertaken by the government or large private companies." He hastened to add that working for a prominent consulting engineer like Mayer would greatly enhance his credentials. The manner in which the position had been offered amused him, and he recounted the story in the same letter.

As with other places where I applied, I came here with the happy prospect that I would be turned away. As is the custom here, I went into the office without knocking, without removing the hat from my head, and said:

"Good morning, Sir; my name is Ammann. I am a civil engineer, have you a job for me?" Finished. Without looking up, Mayer said:

"Well, what have you done?"

Then I told him of my former jobs in brief words, and then he said, "I have no work now, but come again next week."

"Well, good morning, Sir." And I went out the door. In the meantime, I sent him my references and when I came to him today, he said:

"You are all right. I will take you." Then there was a brief discussion about the work, the salary, etc. "Well, you will come Friday at half past eight," he said.

"Well, good morning, Sir," said I.

That is how the Americans are, short and direct without circumspection and when one is also that way, one makes a greater impression than with polite words. The salary is certainly not very impressive, $50 per month, but for the moment, it is secondary for me as long as I have interesting work.[2]

In November 1904, just six months after being hired, Ammann was no less surprised by an American management practice exemplified by Mayer's telling him that the series of bridges he was scheduled to work on would be postponed for a few months. Since there was no work at present, Ammann would not be paid his full salary. Mayer would, however, keep him on at half pay with the promise of restoring his full salary when the project resumed; alternatively, he suggested that Ammann find another job in the interim. Ammann, unwilling to take a reduction in pay, chose the latter and requested a letter of recommendation. Mayer provided a glowing endorsement and a written introduction to several top engineers. In less than a week, Ammann had another job. During his brief tenure with Mayer, he had worked on no fewer than thirty steel bridges and watched with fascination as a team in the office developed the design for a Hudson River crossing at New York City. Hilgard's claim proved true: a young man could indeed quickly gain experience in a wealthy country on the move.

Ammann's second position in America took him to Harrisburg, Pennsylvania. Here he joined a staff of one hundred engineers who supported the Pennsylvania Steel Company's five-thousand-man bridge-building division. Two weeks after he began the job, however,

Mayer attempted to persuade the young man to return to New York. Ammann was unwilling, but he did agree to work for Mayer on the side for an hourly wage.

Ammann was bright, eager, and competent. A self-starter, he could solve any problem he was given, and he proved to be particularly strong in the often neglected area of construction technology. None of his employers wanted to see him go. With each resignation, Ammann was given assurances that a job was waiting for him if he chose to return, and he often went on to provide consulting services to the former employer, maintaining an ongoing association as he moved into new professional realms.

The Pennsylvania Steel Company quickly took notice of this intense, soft-spoken foreigner who chose to spend his lunch hours walking through the workshops observing the company's manufacturing and assembly techniques. Pleased with what he was learning, Ammann wished to extend his stay in the United States, although he was anxious to be reunited with Lilly, and they officially announced their engagement in the spring of 1905. That summer, Ammann took a one-month leave from his job and traveled to Zurich, where he and Lilly were married. After a brief honeymoon in Italy, Mr. and Mrs. Othmar H. Ammann returned to Harrisburg. Seven months later, they were expecting their first child.

AMMANN FORGES A CAREER

During the first months of Lilly's pregnancy, Ammann resigned his position at Pennsylvania Steel, and the couple spent March through August moving three times for three different jobs, in a race to cover as much professional ground as possible before settling down to raise a family. Their first move was to Pittsburgh, where the engineer briefly worked with McClintic/Marshall, an operation similar to the Pennsylvania Steel Company. His job description was comparable with the Harrisburg post, but his title, first engineer, represented a significant career advancement. Two months later, however, he resigned, and the Ammanns moved to Chicago so he could take a job with Ralph Modjeski. Like Joseph Mayer, Modjeski was a highly respected consulting engineer who specialized in long-span steel bridge design. Ammann's senior by eighteen years, Modjeski was a gifted mathematician and artist who turned down an opportunity for a career as a concert pianist to study bridge design, graduating first in his class at the Ecole des Ponts et Chaussées, the esteemed engineering college in Paris. Over the summer, Ammann helped Modjeski's office design a series of steel bridges for the Oregon Trunk Railroad. Many of the spans were quite daring, including a 340-foot-long arch bridge across the Columbia River.

Lilly Ammann explained to her in-laws that Othmar would accomplish two objectives in taking this job: he would be closely associated with the design of long-span structures (his work until then had been confined to short and medium spans), and the prestige of working with Modjeski would prove a valuable bargaining chip when her husband negotiated a new title and salary with the Pennsylvania Steel Company, for the couple intended to return to Harrisburg. Ammann reveled in his work with Modjeski, and he told his parents, "I am continuously occupied with designing and calculating bridges, so there is more responsibility and greater mental effort connected with the work; and with that, more satisfaction and value."[3]

The Ammanns returned to Harrisburg in late August 1906, a month before the birth of their first son, Werner, and rented a small house in the country. Upon his return, Ammann was named one of three first engineers for Pennsylvania Steel, and he was rewarded with a considerably better salary than he had earned before. No longer burdened with drafting and writing construction notes, he now led a design team, performed structural calculations, and prepared cost estimates.

At the time, there were only a handful of acknowledged experts in long-span steel structures. Ammann had already worked with two: Mayer in New York and Modjeski in Chicago. Within a year of being reestablished in Harrisburg, he began a professional association with a third: Frederic C. Kunz of Philadelphia.

Ammann and Kunz met in 1907 after the failure of the Quebec Bridge, one of the most notorious bridge disasters of the early twentieth century. As construction was nearing completion over the St. Lawrence River, the structure turned on itself and came crashing to earth. Horrified by the mountainous pile of twisted rubble, the Canadian government hired Kunz, a specialist in long-span truss design, to investigate the failure. Ammann's employers had completed a contract for the Queensboro Bridge in New York City—a long-span cantilever bridge similar in form to the failed Quebec structure—a few months before the catastrophe. Concerned about the stability of the Queensboro Bridge, Pennsylvania Steel assigned Ammann the responsibility of reviewing and recalculating the design. Kunz was hired by Pennsylvania Steel to consult with Ammann.

Kunz took an immediate liking to the young engineer and soon made a pitch for Ammann to leave his current position in Harrisburg and join Kunz and his partner, Charles C. Schneider, in their Philadelphia-based practice, Kunz & Schneider, Consulting Engineers. Ammann eventually yielded to Kunz's offers, but two years elapsed before he made the move. In the interim, he kept his position with the Pennsylvania Steel Company and spent evenings and weekends assisting Kunz, first with the Quebec investigation (the bridge collapsed because its compression members were too weak), and later with a redesign for the failed bridge. The office of Kunz & Schneider was fairly confident that it would also win the contract to oversee the rebuilding of the Quebec Bridge. Should the work be awarded, Ammann agreed to become Kunz's full-time assistant and move his family to Canada, where he would serve as the project's supervising on-site engineer—an extraordinary opportunity for someone not yet thirty.

Regrettably, Kunz & Schneider did not get the contract; a Canadian firm won the job instead. Kunz still wanted Ammann to work exclusively for his practice, and Ammann wanted the challenge of assisting with the high-profile projects going through the Kunz & Schneider office. So in May 1909 he and his family packed their bags and moved from Harrisburg to a rented house on the outskirts of Philadelphia. There, on the western bank of Wissahickon Creek, Ammann set up a home office and became a full-time consulting engineer to Kunz & Schneider.

This job change continued to advance Ammann along a career path that focused on long-span steel structures. Long-span design was a specialty that supported relatively few practitioners. People at the top did well and enjoyed a high level of esteem among colleagues. The road to success was long, uncertain, and likely to require financial sacrifice. The Ammann family's finances in Harrisburg had been comfortable, thanks to a secure salaried position supplemented with fees earned from evening and weekend work. Being an independent consultant in Philadelphia meant adjusting to a smaller, less secure income. But the nature of the work attracted Ammann, and his family was willing to accept a temporary setback in order to make the new position economically feasible.

With the completion of the Quebec Bridge redesign, Ammann assisted on a variety of projects, including the design of a long-span steel arch bridge across the St. John River in New Brunswick, Canada. He also helped write *Design of Steel Bridges* by Kunz and Schneider, a book that first appeared in 1915 and immediately became a classic among structural engineers; the authors, in the acknowledgements, credited Ammann's substantial contribution.

Although career and family life flourished as the seasons passed along the Wissahickon, it never occurred to Ammann and his wife that they would make this situation permanent. Their long-range plan had always been to return to Europe. With that in mind, Ammann maintained close contact with former teachers, classmates, and employers, periodically arranging opportunities for work in Germany or Switzerland. On several occasions a move was planned but then forestalled in favor of a more seductive possibility in America. The pressure to leave mounted when a second son, George Andrew ("Andy"), was born nine months after the Ammanns moved to Philadelphia. They were keen to have their children educated in Switzerland, and Werner was now old enough to enter kindergarten. So at the close of 1911, the engineer made a determined effort to tie off loose ends with Kunz & Schneider. The family hoped to leave the States in late spring, but the departure was again postponed when another leading American master of long-span design, Gustav Lindenthal, offered Ammann an irresistible job in New York.

Lindenthal, a gregarious Austrian-American bridge builder, had been catapulted to professional stardom during the early 1880s with a series of long-span steel and wrought iron bridges built in Pittsburgh. Lindenthal and Kunz had been good friends long before Kunz met Ammann. Once Ammann and Kunz began their association, an introduction to Lindenthal was inevitable. Ammann and his wife met Mr. and Mrs. Lindenthal at a social event in Philadelphia hosted by Kunz in 1910. The Ammanns and the Lindenthals became fast friends and over the next two years saw each other often. Throughout the period, Lindenthal was designing a colossal railroad bridge—the Hell Gate Bridge in New York—for the Pennsylvania Railroad Company.

Lindenthal was authorized to proceed with construction of the Hell Gate early in 1912. Contract in hand, he asked Ammann to join the project team as inspector of construction. When Ammann hedged, Lindenthal raised the ante and offered his young colleague the position of first assistant to the chief. After Lindenthal, Ammann would be the highest-ranking engineer on the job. Ammann officially accepted the position in April 1912 and began his first job with Lindenthal at the beginning of June.

LINDENTHAL AND AMMANN

For the previous three years Ammann's personal and professional life had been quiet, steady, and far removed from the daily dramas associated with construction. He now found himself shuttling between a monumental building site on the northern end of Manhattan and a bustling project office at the top of a downtown skyscraper on William Street. The network of structures making up the Hell Gate Bridge was awesome: two and a half miles of bridging including viaducts, trestles, two small-span bridges (the Little Hell Gate Railroad Bridge and the Bronx Kill Railroad Bridge), and the network's most conspicuous seg-

1.2 The Hell Gate Bridge during construction. The bridge's chief engineer, Gustav Lindenthal, stands at the center of the group. Ammann, his chief assistant, stands to Lindenthal's right and sports a mustache and felt hat. David B. Steinman is fourth from the left. (Photographer unknown; courtesy of the Library of Congress)

ment, the Hell Gate Arch Bridge. Carrying four pairs of tracks, it hovered above the waters of the East River using Wards and Randall's Islands as stepping-stones between the boroughs of Manhattan, Queens, and the Bronx.

Engineers around the globe were generally fascinated by the scale of the network, but they took particular interest in following the design and construction of the monumental Arch Bridge. A stunning landmark to the Industrial Age's boundless vision and unshakable self-confidence, the structure bristles with strength and energy. Certainly Lindenthal's masterpiece, its 1,017-foot clear span is supported by a steel arch whose eloquent silhouette is drawn with a pair of gracefully curved trusses. When the bridge opened in 1917 it was the longest-spanning arch in the world. Its carrying capacity—seventy-five thousand pounds per linear foot—was unprecedented and has yet to be exceeded.

Throughout their four-year association on the Hell Gate project, Lindenthal and Ammann enjoyed a close rapport. Nearly thirty years Ammann's senior, the well-established Lindenthal was an extrovert whose physical characteristics—tall and portly—served him well. His booming voice and striking white beard made him hard to miss and very much worked to his advantage as he ceaselessly strove to be the center of attention. Ammann, by contrast, was reserved, and only five feet five inches tall. Although he rarely raised his voice, Ammann's formidable intellect and quiet inner strength were thoroughly commanding, and he never had any difficulty holding his own. Lindenthal's skill at self-promotion brought clients to the practice; Ammann's dependability at managing the details assured Lindenthal that his contracts would be fulfilled in a timely, cost-effective manner. Work between them progressed without a hitch until Ammann felt compelled to abruptly leave the Hell Gate project in 1914.

On August 1, Ammann was among a group of Swiss-Americans in New York who had gathered for a picnic to commemorate the founding of the Swiss Federation. Before the festivities concluded, it was announced that the army of Emperor William II of Germany had taken position on the banks of the Rhone River across from Basel; World War I was about to grip Europe. Ammann was still a citizen of Switzerland and a reserve officer in its army. His eldest son, parents, and a brother were currently living in Basel. Patriotism and a concern for family and friends stirred him to immediate action. Ammann booked transatlantic passage before receiving written notice to report for active duty. He left for Switzerland on the morning of August 6, along with two other Swiss engineers who were also working in New York.

While Lieutenant Ammann took command of a small battalion stationed in the St. Gotthard Pass, Lilly Ammann was left alone to manage household affairs during her husband's absence. She couldn't keep up with the rent for their Staten Island house, so she found subtenants for it and moved herself and her younger son to Philadelphia, where they stayed with Othmar's cousins. Werner had been sent to live with his grandparents eleven months earlier so he could attend Swiss public school for one year.

Lindenthal was not pleased with Ammann's abrupt departure. The ever meticulous and thoroughly capable assistant had kept the Hell Gate project running smoothly and on schedule. His impulse for soldiering left Lindenthal with more responsibility than he cared to shoulder. In Ammann's absence, Lindenthal promoted David B. Steinman to the position of first assistant. A precocious mathematician from the Lower East Side, Steinman had been a phenomenon at the City College of New York and Columbia University, receiving three advanced degrees from the latter, including a Ph.D. in engineering at age twenty-four. Outstanding in his own way, Steinman did not have the Ammann touch, and Lindenthal was soon leveraging his personal influence with the Ammann family. He appealed to Lilly and urged her to pressure her husband into suspending his stint of active military duty—Lindenthal was willing to pay whatever was necessary to secure his early discharge.

As it happened, war did not erupt on the Swiss border and Ammann was released from service after three months. Werner returned to America with his father, and the family was reunited in New York on December 11, 1914. Lindenthal immediately reinstalled Ammann as first assistant. It is unclear whether an unhealthy competitive tension existed between Steinman and Ammann before the leave of absence. What is certain is that upon Ammann's return—and Steinman's subsequent demotion—a bitter rivalry took hold that would last throughout their careers as both young men eventually distinguished themselves as gifted and accomplished bridge builders. (It was widely known that the mention of Steinman's name was strictly forbidden in the Ammann household.)

Soon after Ammann's return, Lindenthal won a commission to design a railroad bridge to transport Kentucky coal across the Ohio River to Sciotoville, Ohio, ninety miles southeast of Cincinnati. Lindenthal made Ammann his chief assistant for the Sciotoville Bridge, too. Engineered with a continuous truss, its design was the first to introduce this European-developed structural form to American soil.

The ribbon-cutting ceremony for the Hell Gate Bridge took place on April 3, 1917; the United States entered World War I three days later. The Sciotoville Bridge opened in August 1917.

After more than five years of demanding schedules, Lindenthal's office was now without work. Peacetime projects that were scheduled to start had been indefinitely postponed, and as long as the war persisted the practice was hard pressed. Lindenthal had approached Ammann earlier in 1917 and suggested that they form a partnership when current projects were complete—Ammann, of course, was expected to make an immediate financial investment in founding the firm. Because the business climate was generally unfavorable, Ammann felt it was a poor time to invest what little capital he had in a start-up venture. And his ongoing ambition through the years—to return to Switzerland for employment— was presently untenable as the war dimmed all hope for a civilian job in Europe.

But Ammann urgently needed an income. Lindenthal accommodated by helping him land a job as the manager of a clay mining operation. The Just Such Clay Company in South Amboy, New Jersey, had been losing money for several years. Lindenthal, an investor in the company, convinced the board of directors that Ammann had the wherewithal to turn the business around. It certainly wasn't bridge design, or an opportunity in Switzerland, but it was a challenge and a way to keep the family together. Ammann took the job, left Staten Island, and moved into a grand but decaying estate house that the clay company provided. To the investors' delight, the mining operation soon showed a profit under Ammann's management.

Although he applied himself to the task at hand, Ammann's heart belonged to bridge design. His reputation among engineers had gained considerable luster in the course of the Hell Gate construction. He wrote a remarkably lucid and detailed account of the bridge network's planning and construction that was published by the American Society of Civil Engineers in 1918 and received that year's Rowland Prize, the society's highest publishing honor. At the very moment his star was rising, the thirty-nine-year-old engineer had to content himself with a position in industrial management. Fortunately, hostilities in Europe ended with the armistice of November 11, 1918. The restoration of civil order promised renewed economic prosperity and the opportunity for practice. Ammann once again corresponded with European colleagues, searching for job leads. By 1919 offers had begun to come his way. He sent a letter of resignation to the directors of the clay operation but withdrew it after they argued that they would need at least a year to find a suitable replacement. He resigned again in 1920 with the intention of taking a position offered by the Swiss Federal Railroad Department.

THE HUDSON CHALLENGE

During the three years Ammann persevered in South Amboy, Gustav Lindenthal was drawing up plans to bridge the Hudson. This was not the first time he had involved himself with the challenge posed by this broad river. Between 1885 and 1903 Lindenthal had designed and partially constructed a suspension bridge engineered to carry twelve rail lines between Hoboken, New Jersey, and West Twenty-third Street in New York. The project was sponsored by a fractious syndicate named the North River Bridge Company whose investors were a group of privately owned railroads, including the Pennsylvania Railroad, the syndicate's most influential and steadfast member. All the sponsors were anxious to reap commercial benefits from the bridge, and with great ceremony ground was broken in June 1895. Soon thereafter dissent and defection among members caused lengthy delays. When new technologies at the turn of the century lowered the cost of tunneling, the North River Bridge Company abandoned its project altogether. In 1903, the Pennsylvania Railroad began a tunnel between New Jersey and a monumental station to be built in Manhattan at

1.3 Gustav Lindenthal's 1921 proposal to bridge the Hudson River at West Fifty-seventh Street in Manhattan. Ammann worked under Lindenthal on the project's development and promotion. (Courtesy of *Scientific American*)

West Thirty-fourth Street and Seventh Avenue: Pennsylvania Station, designed by architects McKim, Mead, and White. Station and tunnel were completed in 1911.

The president of the Pennsylvania Railroad, Samuel Rea, had been an enthusiastic Lindenthal supporter through the North River Bridge Company and continued his support after the syndicate's disbanding. When the Pennsylvania Railroad began plans for a bridge complex over the Hell Gate that would connect the northern branch of their coastal line to Manhattan, Rea made certain that the design commission was steered in Lindenthal's direction. Unlike the Hudson project, financing and construction went smoothly. Ammann, of course, had been a factor in the Hell Gate's success.

Rea and Lindenthal never abandoned their dream to make history by bridging the Hudson. Wealthy though it was, the Pennsylvania Railroad could not sponsor the project alone, so soon after the Hell Gate opened Rea encouraged Lindenthal to work with him on reviving the North River Bridge Company. In a matter of months a new group had formed.

It had been more than thirty years since Lindenthal proposed his first Hudson River crossing. The 1888 design had been nothing short of audacious; his suspension bridge proposal of 1920 was even bolder. Connecting West Fifty-seventh Street in New York with Fiftieth Street in Weehawken, New Jersey, the bridge comprised a double-decked road system, with an upper deck that accommodated pedestrian walkways and sixteen vehicular traffic lanes and a lower deck that supported twelve railroad lines. Its clear span was more than a half mile long—3,240 feet. Its 840-foot steel and granite towers soared higher than the day's tallest skyscraper, the Woolworth Building in New York City. (Lindenthal and the new North River Bridge Company were so confident in their power and so attuned to grandeur that it seemed only fitting to at once build both the world's tallest towers and the world's longest bridge.) Lindenthal estimated that the project would cost $100 million, but he also figured that a profitable return would soon be realized from tolls, with additional income generated by rent revenues from an office tower built on top of the Manhattan anchorage. The members of the resurrected North River Bridge Company were convinced the scheme was sound business.

Lindenthal's design gained wide notoriety through its publication in the April 1921 issue of *Scientific American*. Predictably, responses to it varied. Many heralded the structure as an engineering masterpiece and the long-awaited solution to escalating traffic congestion at ferry boat landings. Others were skeptical of the proposal's genuine feasibility. Powerful members of the Manhattan business community in the West Fifty-seventh Street area who were slated to be displaced by the structure were mortified. New Yorkers fortunate enough to be spared displacement were nonetheless enraged by the prospect of tens of thousands of additional automobiles, buses, and trucks circulating through the already choked streets of their neighborhood.

Ammann returned to Lindenthal's office four months before the proposal appeared in the magazine. Convincing Ammann to join him had taken some doing on Lindenthal's part. A long-span bridge of this magnitude was bound to capture Ammann's interest, but at forty-one concerns over his economic security gave him pause. As an enticement, Lindenthal promised to form a partnership with him once the North River Bridge Company's project was under construction. Lindenthal even offered Ammann land on his estate in Metuchen, New Jersey, and encouraged his future partner to build a house there. Ammann informed the Swiss federal railroad that he would not be joining their staff as planned, and he accepted the position with Lindenthal but declined the land offer. Instead, in July 1921, he purchased a rambling house in Boonton, New Jersey—a handyman's special—with a down payment lent by Lindenthal.

Ammann's first months back in Lindenthal's office were consumed with planning the gargantuan Hudson River suspension bridge, but by the end of the year work on the project had slowed. During most of 1922 and 1923, therefore, Ammann assisted on a three-bridge commission Lindenthal had won for the Willamette River in Portland, Oregon: the Ross Island, a cantilever bridge, the Burnside, a truss bridge, and the Sellwood, a cantilever bridge. These contracts brought substantial fees to the practice. Ammann, who managed the Oregon work, approached contract fulfillment with a mixed response: he was concerned to see the practice's stream of commercial work dry up, but he was also pleased to be able to once again focus his attention on the Hudson River bridge project. His future partnership with Lindenthal, after all, was directly tied to its success.

By the time the Oregon work was complete, the Hudson design was fully developed and Lindenthal's staff busied itself refining traffic studies and cost estimates. Ammann was often asked to present the scheme at public and professional meetings. The discouraging sentiments consistently voiced from the audience gradually convinced him that Lindenthal's project—with its twenty-eight lanes of traffic and $100 million price tag—was simply too overreaching in its scale and cost. Furthermore, the site of the proposed bridge—particularly at its Manhattan connection—remained saddled with seemingly intractable difficulties pertaining to real estate and easement acquisition. Indifferent to public opinion, Lindenthal and his powerful cronies in the railroad industry also seemed blind to much of the social change that followed World War I. The great era of the railroad—which reached its peak when these men were in midcareer—was being overtaken by the era of motor vehicles: private, public, and commercial. They couldn't see it—or didn't wish to acknowledge it—but the demand for rail service was in permanent decline.

A MASTER BUILDER EMERGES

As opposition to the North River Bridge Company scheme gained force, Lindenthal withdrew into a world of militant self-assurance. Ammann felt certain that unless there was compromise on the location of the bridge and the number of lanes it accommodated, the project would be killed and the Lindenthal office dismantled. Ammann wanted a successful project. Success meant continued practice in long-span bridge design and a professional partnership. Ambitions were on the line. When he diplomatically challenged Lindenthal, he was quickly silenced. After the confrontation, Ammann recorded, "G. L. [Lindenthal] rebuked me for my 'timidity' and 'shortsightedness' in not looking far enough ahead. He stated that he was looking ahead for 1,000 years."[4] Lindenthal continued to refuse to alter his plan, neither reducing the scale of the project nor reconsidering its site. All his employees began to awaken to the sobering realization that their ship was sinking.

Ammann now faced the most difficult situation of his professional life. The celebrated engineer who had been his trusted mentor was hell-bent on pursuing a course certain to lead to disappointment and financial disaster. After three years of employment in Lindenthal's Dey Street office, Ammann found his household in deep economic turmoil, partly because six months earlier he had agreed to a temporary salary reduction to help keep the office afloat. The family nest egg, painstakingly built during the Harrisburg years, had been chipped away to keep things going while Ammann gambled on a partnership with Lindenthal. His predicament was forcing him to face a difficult choice: he could stay in America, leave Lindenthal's office, abandon long-span design, and join his profession's rank and file; or he could at long last move his family to Switzerland and take a position there. He also envisioned a possibility that would be his riskiest career move ever. The characteristically conservative Ammann, with little left to lose, undertook one more gamble: the engineer struck out on his own.

Forty-four years old, on March 21, 1923, Ammann left the employ of Lindenthal and set up an office in a loft building at 470 Fourth Avenue in New York. There he developed his own suspension scheme to bridge the Hudson.

During his first six months of self-employment, the engineer kept his break from

Lindenthal—and the anguish that he lived with prior to it—a secret from family members in Switzerland. His father had died six years earlier, but in a Christmas letter dated December 14, 1923, Ammann explained his situation to his mother.

Christmas has to be approaching for me to get around to pick up a pen and write to you after so many months. You would be justified in thinking that I am an ungrateful son. Nonetheless, I should like to assure you, and perhaps you felt ringing in your ears, that my thoughts were so often with you. The quietness and patience that are so necessary for bringing one's thoughts on paper were lacking.

In order for you to understand my situation for many months, in fact for the whole year, I will no longer conceal from you that the giant project for which I have been sacrificing time and money for the past three years, today lies in ruin. In vain, I as well as others, have been fighting against the unlimited ambition of a genius that is obsessed with illusions of grandeur. He has the power in his hands and refuses to bring moderation into his gigantic plan. Instead, his illusions lead him to enlarge his plans more and more, until he has reached the unheard of sum of half a billion dollars—an impossibility even in America.

However, I have gained a rich experience and have decided to build anew on the ruins with fresh hope and courage—and, at that, on my own initiative and with my own plans, on a more moderate scale. It is a hard battle that I have already been fighting for six months now, but the possibility of success is constantly increasing, so that I do not allow myself to be frightened in spite of the great handicaps and my shrinking finances. I wait and hope that the New Year will finally bring my work to fruition.[5]

The proposal Ammann developed for bridging the Hudson was officially unveiled at a meeting of the Connecticut Society of Engineers on February 19, 1924. By the degree to which Lindenthal's scheme seemed overambitious and overblown, Ammann's seemed disarmingly restrained and eminently doable. The younger engineer's proposal emphasized vehicular traffic, envisioning a wide roadway that would accommodate eight lanes of traffic and two pedestrian walkways on the upper deck, and four light-rail lines on a lower deck that would be constructed in the future when capacity was reached on the original deck. The estimated price tag was a modest $40 million (the estimate for Lindenthal's structure had grown by this time to $500 million); and, with a location at the northern end of Manhattan connecting with a sparsely populated section of Bergen County, New Jersey, Ammann's site avoided the pitfalls of developing approaches and interchanges on land where

1.4 Ammann's proposal for a crossing between Fort Lee and Fort Washington as first presented to the public in 1924. (Rendered by Othmar H. Ammann, 1923; courtesy of Margot Ammann-Durrer)

real estate prices—and emotions—ran high. The appearance of the bridge—with its 3,500-foot clear span—was a vision of simplicity and grace. Eventually named in honor of the first United States president, Ammann's George Washington Bridge became a landmark achievement in the development of suspension technique.

Ammann's suspension bridge drew on a different design tradition than did Lindenthal's. Lindenthal's structure relied on deep chain trusses strung between its towers to prevent excessive motion on the road deck. He had invented this suspension technique and first used it in his vehicular crossing in Pittsburgh, the Seventh Avenue Bridge, in 1884. Though never attempted for a span of its length, reputable engineers agreed that as long as the individual members of the Hudson bridge were strong enough (that is, massive enough) the bridge would stand.

Ammann's design represented a stunning—though logical—breakthrough in an approach that reached back to the form's origins: engineer road deck stability through the design of the road deck itself, as had been done with the earliest vehicular suspension bridges.

Short-span suspended foot bridges have been built since ancient times, but an American, James Finley, was the first designer to demonstrate that suspension bridges could be configured with plank decks to accommodate vehicular traffic. Between 1801 and 1828 he erected a series of thirteen suspension bridges, all of which had level floors and were intended for wagons and carriages as well as pedestrian traffic. The first of these crossed Jacob's Creek in Uniontown, Pennsylvania, with a clear span of 69 feet. His second bridge—a 128-foot span over the Potomac River—opened in 1804. In 1808 Finley both patented his suspension system and presided over the opening of a 306-foot suspended span he designed to bridge the Schuylkill Falls near Philadelphia. These structures possessed the same elements as Ammann's bridge: metal rods or cables connected horizontal decks to the main suspension members of metal chains or bundled wire, which were draped over towers and anchored to the ground.

Finley's were remarkably fine structures for their day, especially because the physics of stress were virtually unknown in engineering circles (progress was based exclusively on intuition and trial and error), and because the materials, methods of fabrication, and construction details were crude. Though several of Finley's bridges ultimately failed—one under the strain of a drove of cattle, another from accumulated ice and snow—the engineer fully appreciated the importance of road deck rigidity, historically the most problematic feature of a suspension bridge. Finley relied for this rigidity on stiff wooden floors made with massive timbers. His 1809 Merrimack River Bridge at Newburyport, Massachusetts, lasted a century and was replaced in 1909 with a bridge that closely follows the original design.

The main arena for pioneering work in suspension design shifted to Europe after Finley's death. Thomas Telford's Menai Strait Bridge was a huge success when it opened in 1826. Built at Bangor in North Wales with a 580-foot clear span, it is generally considered the world's first long-span suspension bridge. Telford followed Finley's example and hung his roadway from iron chains, but he did not use a heavy wooden deck to counteract the structure's tendency to sway and deform excessively. Instead, Telford specified a lightweight timber deck and stabilized it with a secondary network of inclined cables in the form of chains running from the towers to the deck. Elegant, strong, and economical, the bridge inspired European engineers on both sides of the English Channel to experiment with the form.

Despite the success of the Menai Strait Bridge, numerous early suspension bridges in Europe failed. For the sake of economy, designers had pushed to make their structures as light as possible. As a result, too often road deck rigidity was inadequately addressed. Most of the failures occurred when suspended decks collapsed after being thrown into motion by wind or traffic. The problem became so pervasive that the French government for twenty years outlawed suspension bridge construction.

While enthusiasm for suspension declined in Europe, interest was on the rise in the United States owing largely to Charles Ellet's brilliant work based on French prototypes. The first American native with a European education in engineering—he attended the Ecole Polytechnique in Paris—Ellet's Schuylkill River Bridge of 1842 in Philadelphia was an elegant demonstration of the suspension form's aesthetic and practical potential. It was

1.5 The Menai Strait Bridge, designed by Thomas Telford, opened in 1826. It is widely considered the first long-span suspension bridge. (Delineated by W. H. Barflett and J. Rogers; courtesy of the Avery Library, Columbia University)

1.6 The Schuylkill River Bridge in Philadelphia, designed by Charles Ellet, Jr., opened in 1842. (Delineated by W. Coome; courtesy of the Avery Library, Columbia University)

the first wire-cable—as opposed to chain—suspension bridge in the New World, incorporating ten cables in its 358-foot span. Ellet's 1,010-foot bridge over the Ohio River in Wheeling, West Virginia, opened seven years later and nearly doubled the clear-span length of the Menai Strait Bridge. In both, Ellet attached continuous deep trusses along the edge of the road deck to minimize deflection.

John A. Roebling, four years older than Ellet, emerged as the world's leading proponent and innovator of suspension bridging in the latter part of the nineteenth century. Roebling was a German native who had immigrated to the United States in 1831. Like Ammann, he arrived in his early twenties with an outstanding education (he had read philosophy under Hegel and studied architecture, bridge construction, and hydraulics at the Polytechnic Institute in Berlin). In 1841, he began manufacturing iron wire. The wires and cables he produced quickly replaced fiber rope in transportation, construction, and agriculture, and Roebling amassed a fortune.

His first bridge was a suspended aqueduct of modest span that carried the Pennsylvania Canal over the Allegheny River. Opening in 1845, the clear span was supported by two iron cables attached to their anchorages with eyebars embedded in masonry. Eyebars are long, flat metal plates with holes punched at either end. Through the "eye" above ground were threaded the wires composing the suspension cable. The buried eye was anchored to the foundation before being embedded. Roebling invented the eyebar for this structure, and it became a standard fastening technique among engineers—one that Ammann used in all his suspension bridges.

Roebling established himself as a master builder with the opening of the Niagara Railroad Bridge in 1859. Ellet had started the project but resigned the commission over a dispute with the sponsors soon after the temporary footbridge was finished. Roebling redesigned and built the double-deck suspended structure that was the first to carry railroads—trains were confined to the upper deck, vehicles and pedestrians to the lower level. The traveling public marveled but often crossed with trepidation. As Mark Twain described a trip across the bridge: "You drive over the Suspension Bridge and divide your misery between the chances of smashing down two hundred feet into the river below, and the chances of having a railroad-train over head smashing down onto you. Either possibility is discomforting taken by itself, but, mixed together, they amount in the aggregate to positive unhappiness."[6]

In 1857, Roebling's son, Washington, joined his father in practice and together they successfully completed numerous long-span suspension bridges, all of which incorporated wire cable, eyebars, and road deck stiffening trusses from anchorage to anchorage. Their crowing achievement, the Brooklyn Bridge across the East River in New York City, was deservedly the nineteenth century's most celebrated bridge of any type. Its scale, majesty, and traffic capacity were unprecedented. Completed in 1883, it boasted a 1,595 1/2-foot clear span (535 1/2 feet longer that the previous record holder, the Roeblings' 1866 Cincinnati Suspension Bridge over the Ohio River). The Brooklyn Bridge advanced numerous innovations. The underwater construction of the tower foundations using the caisson method 78 feet below the surface of the water had never before been applied at these depths or for a structure of this size. It was the first bridge to use steel wire for the cables, which proved to be 75 percent stronger than the wrought iron that had previously been used. For the first time, the wires were protected against rust with a galvanized coating, which has helped preserve them to this day. The method of spinning cables with thin wires strung one wire at a time with a wheel that traveled back and forth from one anchorage and tower to the other had previously been devised by Roebling for the Niagara Railroad Suspension Bridge, but the technique's efficiency and speed were fully perfected in the course of the Brooklyn project.

The East River in New York soon gave rise to two other innovative and exceptionally long suspension bridges. The Williamsburg Bridge, an ungainly and graceless hybrid designed by Leffert L. Buck, opened to traffic on December 19, 1903, with a 1,600-foot clear span (four and a half feet longer than that of the Brooklyn Bridge). It was the first suspension bridge with all-steel towers, and for this reason alone it holds a place in the history of technology. Though shorter in length (1,470 feet), the Manhattan Bridge, which opened in 1909, is a far more significant structure. It provided the technical and aesthetic stepping-stone between suspension design in the nineteenth century and the great advancements of the twentieth century. Designed by Leon S. Moisseiff, the all-steel Manhattan Bridge's simple, restrained appearance was derived from the innovative design theory that engineered it.

With the Manhattan Bridge, Moisseiff—a brilliant mathematician of Latvian origin—introduced the "deflection theory" to America. The theory was formulated in Austria by Joseph Melan, an expert in reinforced concrete arches. Moisseiff developed the theory's principles, applied them to long-span suspension design, and in so doing gave rise to all the slender bridges that proceeded. Deflection theory holds that as the deadweight of a suspension bridge increases per linear foot, the need for deck stiffness decreases, largely because the gravitational pull on monumental suspension cables, suspender cables, and unstiffened decks alone is nearly sufficient to provide a level of resistance against the force of wind and moving traffic, eliminating the need for excessive add-on devices like stiffening trusses or cable stays. (It is a principle that harkens back to Finley's use of heavy timber planking, but for a scale of bridge he could not have imagined.) Moisseiff translated this phenomenon into a series of mathematical formulas essential to the theory's rational application. The equation's overall impact on form, exemplified by the Manhattan Bridge, was profound. Every aspect of the bridge—even the towers—is delicate and made aerial in appearance by the reduction of hundreds of tons of iron and steel that otherwise would

1.9 John A. and Washington Roebling's Brooklyn Bridge opened in 1883 (several years after John Roebling's death). The massive granite towers of the bridge support four bundled steel wire cables. Iron stiffening trusses running along the suspended roadway from anchorage to anchorage stabilize the structure, as do inclined stays running diagonally from the towers to the deck. (Photographer unknown; courtesy of the New York Pubic Library)

have been required by established design practice. Considerable cost savings accompanied this reduction, making the bridging of previously unbridgeable spans (such as the Hudson River) genuinely economical for the first time.

Ammann's design for the Hudson River would be the first to rely on deflection theory for a suspension bridge of such great length—its clear span more than twice as long as the Manhattan Bridge's. Improvements in the quality and strength of materials contributed to

1.10 The Manhattan Bridge, designed by Leon S. Moisseiff, opened in 1909. (Photo, George P. Hall; © Collection of the New-York Historical Society)

the engineering, but Ammann's innovative application of the deflection theory was the most significant force in determining the bridge's form. The boldness of the design is magnified by the fact that the Manhattan Bridge had the stabilizing benefit of a double-deck span tied together with stiffening trusses. Using neither Lindenthal's chain suspension trusses nor Ellet's, Roebling's, and Moisseiff's deck trusses, Ammann enhanced the rigidity of his road deck structure with a pair of deep girders to give the deck its ribbonlike appearance.

1.11
The George Washington Bridge
at West 178th Street in Manhattan
opened to the public on October
25, 1931. (Photographer unknown;
courtesy of the Library of
Congress)

PATRONAGE FROM THE GOVERNOR'S MANSION

Unlike Lindenthal and his North River Bridge Company, Ammann's Hudson River proposal initially lacked significant backing from powerful business and political interests. His efforts to promote the scheme entailed a seemingly endless number of presentations to engineering societies, small business groups, political organizations, and community associations. Not only was his scheme in competition with Lindenthal's, several businesses and engineering concerns were lobbying the federal government with tunnel proposals. Each project had its merits, but one thing was certain: political will and commercial interests would muscle the prevailing scheme to construction by the end of the decade.

Ammann did have one political ally in high office from the beginning: Governor George S. Silzer of New Jersey. There is little doubt that his assurance of support was a significant factor in bolstering Ammann's courage to break from Lindenthal. At first the governor kept his support from public view while he quietly coached and directed Ammann's efforts. Part of Silzer's motive in maintaining a low profile was strictly personal: He and Lindenthal had been friends and business associates for years. Both were investors in the Just Such Clay mine, and it was through that mutual connection that Silzer first become acquainted with Ammann. Although the former state senator had once backed Lindenthal and the North River Bridge Company, after he assumed the New Jersey governorship in 1923 his strong and sustained support was discontinued.

Silzer was a Democrat who modeled his policies after those of Woodrow Wilson–style liberals—during Wilson's term as governor of New Jersey, in 1911–1913, Silzer had been one of his chief aides. Like Wilson, Silzer strongly advocated public works and was particularly eager to attach his name to popular highway and bridge projects. The legislative atmosphere in which he operated as governor was dominated by Republicans who for the most part opposed Silzer's initiatives simply because he was a Democrat. His political future became tied to promoting projects that would squarely benefit Republicans and therefore win their support. The Lindenthal scheme held no particular benefit for New Jersey Republicans, and by 1923 it was clear that the North River Bridge Company's project had lost favor with the public at large. By contrast, Ammann's bridge proposal struck Silzer as an easy mark for broad popular support. As for the governor's political opponents, Ammann's bridge connected New York to Bergen County, a Republican stronghold and largely rural district that would be opened to rapid economic development with the construction of the bridge: this solution to the Hudson challenge appealed to even the staunchest New Jersey Republicans.

Before Silzer publicly announced his endorsement of the Ammann proposal, he felt obliged to inform Lindenthal. To ease the way, Silzer asked his longtime colleague to review a copy of Ammann's prospectus. Lindenthal's response was bitter:

> Mr. Ammann has been my trusted assistant and friend for ten years, trained up in my office and acquainted with all my papers and methods. But I know his limitations. He never was necessary or indispensable to me. . . . Now it appears that Ammann used his position of trust, the knowledge acquired in my service and the data and records in my office, to compete with me in plans for a bridge over the Hudson and to discredit my work on which I had employed him. He does not seem to see that his action is unethical and dishonorable.[7]

Silzer's support of the Ammann scheme would leave no permanent rift between him and Lindenthal. Ammann's decision to compete against Lindenthal, however, forever changed the senior engineer's regard for his colleague. Their friendship was never reestablished; Lindenthal, until his death in 1935 at age eighty-five, continued to unsuccessfully promote his Hudson River crossing as president and chief engineer of the North River Bridge Company, an enterprise that came to exist largely in name alone.

As for the ongoing collaboration between Silzer and Ammann on the George Washington Bridge, once the governor made a public endorsement a domino effect followed, with politicians, business leaders, and newspapers throwing their weight behind a vehicular suspension bridge between Fort Washington in New York and Fort Lee in New Jersey. The legislatures of both states gave formal approval to the Ammann scheme in March 1925, entrusting its construction and operation to the recently formed Port of New York Authority (later called the New York and New Jersey Port Authority, and later still the Port Authority of New York and New Jersey). Ammann's bridge across the Hudson was the first facility that the authority was instructed to build.

In the short term, Ammann faded from the process. His original and carefully considered proposal had proved to be the perfect fit between genuine need and the political realities of the day. He had independently circulated his design in the public domain with no guarantee that, if his vision became public policy, he would himself be given a role to play in its development. The public had come to enthusiastically embrace his plan, the Port of New York Authority was about to build it, and Ammann found himself sidelined, jobless and without a mission. But not for long. Silzer remedied the situation within months of the project's official approval by encouraging the authority to establish the position of bridge engineer and to name Ammann to the post. The engineer joined the staff of the Port Authority in July 1925 and became a United States citizen later that year.

CIVIL SERVICE, 1925–1939

Ammann's appointment made him the chief of a small professional staff reinforced with several consulting engineers. While he and his team worked out all the final details for the George Washington Bridge, the state legislatures of New York and New Jersey passed resolutions authorizing the Port Authority to study and prepare plans for bridging the Kill Van Kull separating Bayonne, New Jersey, from Port Richmond, Staten Island. As the Port Authority's engineer of bridges, Ammann designed the Bayonne Bridge in 1926–1927. Construction of the George Washington Bridge began in 1927, and ground was broken for the Bayonne Bridge in 1928. Between 1928 and 1931 Ammann oversaw construction of both the world's longest suspension bridge (the George Washington) and the world's longest arch bridge (the Bayonne). Both bridges opened in 1931. By all accounts, Ammann was now an established master of long-span steel.

During his first three years with the Port Authority, Ammann supervised the construction of two additional crossings. Both had been commissioned before his tenure, both were truss bridges, and both had been designed by J. A. L. Waddell of Waddell and Hardesty, Consulting Engineers. The Goethals and Outerbridge Crossing Bridges joined Staten Island and New Jersey at sites south of the Bayonne Bridge.

In 1931, Ammann began plans for the first tube of the Port Authority's Lincoln Tunnel under the Hudson River, connecting Weehawken with midtown Manhattan. Construction commenced in 1934. That same year, he began his association with Robert Moses, the visionary planner and builder who would become New York's most successful promoter and administrator of highway projects.

In 1933, Moses was named chairman of the newly created Triborough Bridge Authority. The authority had been established to complete an ambitious bridge project across the East River that the city of New York had begun building in 1921 but abandoned in 1929. Like the Hell Gate Bridge, the Triborough Bridge was not a single structure. Rather, it was a network comprising three long-span bridges—a suspension bridge, a standard truss bridge, a truss bridge with a movable road deck—and three and a half miles of viaduct.

The only parts of the bridge completed under the city's administration were the anchorages and tower foundations for the suspension span. Moses persuaded Ammann to join the authority as its chief engineer, supervising the creation of an engineering department, while concurrently serving in the same capacity for the Port Authority. As chief engineer, Ammann oversaw the construction of the entire Triborough Bridge network and redesigned all its structures except the Triborough Truss Bridge, which was designed by Ash-Howard-Needles and Tammen, Consulting Engineers. The project was executed with remarkable speed: Ammann assumed responsibility for the work in 1934, and the bridge was opened in 1936.

Ammann began designing the Bronx–Whitestone Bridge for the Triborough Bridge Authority in 1935. This was the first long-span suspension bridge he designed from scratch since the George Washington. The Bronx–Whitestone as designed and constructed was the most ethereal, diagrammatic suspension bridge of the century. Its monolithic towers were the first ever to be made with plate steel. The simple, fine line of the road deck was the result of the girder plate engineering devised for the George Washington Bridge. Ammann and his team designed and constructed the Bronx–Whitestone in time for the motoring public to cross it on their way to the New York World's Fair of 1939 in Flushing Meadows, Queens.

Before the World's Fair officially closed in October 1940, the Queens–Midtown Tunnel gave visitors—and residents—their first opportunity to drive between the boroughs of Queens and Manhattan by crossing under the East River instead of over it. Moses championed the tunnel, thereby expanding the Triborough Bridge Authority's domain; its name, accordingly, was changed to the Triborough Bridge and Tunnel Authority in 1938. Ammann did not design the Queens–Midtown Tunnel, but he did supervise the completion of its construction.

The flurry of New York bridge and tunnel construction that began in the late 1920s with the George Washington Bridge continued intermittently for more than thirty years. This unprecedented pace of construction was integral in lacing the New York region together in patterns that would maintain its prominence as a major hub along the Atlantic Coast's northeast corridor. The network that united the region with highways, bridges, and tunnels was built largely as envisioned in 1929 by the Regional Plan Association, a nonprofit group founded in 1914. Through his various public offices, Moses got much of the ambitious network built: no small feat considering the project's enormous cost and that much of it was financed during the Great Depression. Economic resources in the New York region were woefully insufficient during the 1930s and 1940s. With the help of his allies in Washington, Moses siphoned huge federal appropriations into his highway projects from such programs as the Works Progress Administration, shrewdly combining federal aid with revenues from the sale of transportation bonds backed by future tolls from the bridges and tunnels.

Ammann's strategically sited Hudson crossing forged the pivotal link that put the construction of the region's vast highway system in motion. It also established his career as a bridge designer. It was his good fortune that his renown as an engineer of efficient, beautiful long-span bridges corresponded with an urgent need for them in his own backyard, as his first clients—the Port Authority of New York and New Jersey and the Triborough Bridge and Tunnel Authority—became two of the country's most active long-span bridge commissioners between 1925 and 1965. Ammann proved to be the right man in the right place at the right time.

His emergence as a designer surprised many of his colleagues. Prior to his break from Lindenthal, Ammann had always been the thoroughly dependable—even brilliant—workhorse who managed all the details. Such had been his role with all the great bridge designers under whom he had worked: Mayer, Modjeski, Kunz, and Lindenthal. He had never had a hand in the conceptional stages of these masters' designs. In fact, there is no evidence from journals, letters, office papers, or anecdotes that, prior to his Hudson proposal, Ammann had wrestled with an ambition to be a designer or to put ideas for hypothetical structures on paper. (He didn't have a Hudson proposal ready when he left Lindenthal—though it would have been wise if he had.) Once he began, each project that the relatively late-blooming Ammann designed and managed resulted in a stunning structure that was completed on schedule and within budget (several actually opened ahead of schedule and under budget). His clients were justified in placing in him their absolute trust.

While his career advanced seamlessly during the 1930s, Ammann's personal life included tragedy. His wife, Lilly, was diagnosed with breast cancer early in 1930; she underwent surgery and soon recovered, but the cancer returned. After two years of diminishing strength, she died on December 18, 1933. Ammann had known and loved her since their childhood. They were the parents of three children, one of them still a dependent. Together they had struggled to build an engineering career that had recently flourished beyond reasonable expectation. This was supposed to be their time together in the glow of success.

During his first summer as a widower, Ammann traveled to the West Coast to serve on the board of engineers for the Golden Gate Bridge. While there, he decided to drive from

San Francisco to Los Angeles to make a condolence call on the recently widowed wife of a longtime colleague and fellow Swiss-American engineer, Fredi Noetzli. Ten years younger than Ammann, Noetzli had also graduated from the Swiss Federal Polytechnic Institute and had been a member of Ammann's fraternity. Both had come to America soon after graduation (Ammann in 1904, Noetzli in 1915). Noetzli eventually settled in California, where he became a noted designer of reinforced concrete dams and was awarded a patent for a form of inclined joint that came to be widely used in their construction. After he died in 1933 at age forty-one, while his estate was being settled, an engineering associate challenged his widow's exclusive claim to the patent. Several engineers sympathetic to her asked Ammann to help with the defense. Ammann interceded, and the lawsuit was quickly resolved in Mrs. Noetzli's favor. A visit from the New York engineer now seemed in order, especially since he, too, had recently lost a spouse.

They had met only once before, when the future Mrs. Noetzli (née Klary Vogt) had traveled to New York from Switzerland to join her fiancé in America. On the day she arrived in 1917, they exchanged their marriage vows at City Hall in Manhattan. Afterward, husband and wife walked three blocks across town to greet Ammann at Lindenthal's William Street office, where he was assisting with the Hell Gate Bridge project. Ammann treated the couple to a celebratory lunch. In the years that followed, Ammann often exchanged letters with the engineer, but he had no further contact with his colleague's wife.

The reunion took place in Mrs. Noetzli's suburban home. It was a brief meeting with a respectful farewell. Ammann drove back to San Francisco and then returned to New York.

When he flew to the West Coast again in November for another meeting with the Golden Gate Bridge Authority, he paid a second visit to Mrs. Noetzli's home. This time, after dinner, Ammann asked his hostess to marry him. Understandably she was taken by surprise, but on reflection she consented. They were married in March 1935. At the time, Ammann's children were grown or nearly so. Werner had followed in his father's footsteps and become a civil engineer. Andy was in graduate school and would later distinguish himself as an ornithologist and outdoorsman. Margaret (Margot), Ammann's youngest child and only daughter, had been born in 1922, soon after the engineer had begun work on Lindenthal's second Hudson River crossing. Margot would enter college, graduate from medical school, and establish a clinical practice in New York City.

THE CONTRIBUTIONS OF A SEASONED MASTER

By 1939, Ammann had been the Port Authority's chief engineer for fourteen years, overseeing the growth of its engineering division from a skeletal office to an enormous support staff with hundreds of employees. The chairman of the authority intended to "promote" Ammann, now sixty, to an executive position that would remove him from the daily workings and management of the authority's projects and staff. Several of the other senior engineers were also tapped for promotion. Ammann had no intention of abdicating his hands-on involvement with design and construction. He and Klary went to Europe for a three-month holiday. When they returned, Ammann, along with five colleagues, resigned from the Port Authority and formed a private consulting practice, Othmar H. Ammann & C. C. Combs, Consulting Engineers. Ammann concurrently resigned his position as chief engineer for the Triborough Bridge and Tunnel Authority. Both authorities, however, would continue to use his services.

Ammann & Combs stayed busy. Founding partner and landscape architect Charles C. Combs attracted numerous highway contracts that collectively formed the backbone of the consultants' practice during their first years together. The projects Ammann brought to the firm included consultation on the design of the Delaware Memorial Suspension Bridge over the Delaware River at Wilmington; plan development for a suspension bridge over the York River in Yorktown, New York; and two contracts for structures in New York City.

The first New York contract came within weeks of the practice's establishment. The Triborough Bridge and Tunnel Authority needed a pedestrian lift-bridge to connect East 103d Street in Manhattan with recreational facilities on Wards Island. The bridge would cross a navigable channel in the Harlem River; it therefore needed to be designed with a lift mechanism that raised and lowered the height of the clear-span deck. Christened the Harlem River Pedestrian Bridge (it did not open until 1949), it has the shortest span among Ammann's New York bridges, but it was one of the engineer's favorite designs—he affectionately referred to it as his "Little Green Bridge."

The city of New York asked Ammann & Combs in 1941 to oversee a thorough inspection of the Brooklyn Bridge. The engineer found the nineteenth century's most technically advanced suspension bridge in remarkably fine shape, even though modern testing techniques revealed that the original building materials and assembly techniques were uneven in quality. The alterations and reinforcements Ammann specified bolstered the structure and preserved it well into the future.

Ammann's most significant contribution to the engineering profession during his initial return to private practice began with a headline-grabbing bridge failure near Tacoma, Washington. On November 7, 1940, the deck of the Tacoma Narrows Bridge over Puget Sound began to oscillate in the wind. Though the wind remained steady, the oscillations grew from gentle sways to violent heaves. The structure's road deck was unable to absorb the tremendous torsional stress caused by the twisting and snapping, and destruction followed. Clark H. Eldridge, the on-site bridge engineer, provided the following eyewitness account.

At about ten o'clock Mr. Walter Miles called from his office to come and look at the bridge, that it was about to go. This was the first indication I had that anything of an unusual or serious nature was occurring.

I immediately drove with Mr. Miles to the dock, from which we could see the bridge. The center span was swaying wildly, it being possible first to see the entire bottom side as it swung into a semi-vertical position and then the entire roadway.

It was at once apparent that instead of the cables in the main span rising and falling together, they were moving in opposite directions, thereby tilting the deck from side to side. I could observe one car, stationary, some distance east from the center. It appeared that the center of the span was remaining about horizontal and the two halves were revolving about a longitudinal axis of the bridge.

. . . Then I observed on the main span that the concrete sidewalk around the stiffeners of the girders was failing badly. The curbs at the construction joints were also failing. Adjacent to the girders, it appeared that the concrete was entirely free and the girders and the concrete were working back and forth continuously three or four inches. The concrete roadway showed no signs of cracking. The main span was rolling wildly. . . . The wind had not moved it a noticeable amount sidewise. The deck, however, was tipping from the horizontal to an angle approaching forty-five degrees.

. . . The entire main span appeared to be twisting about a neutral point at the center of the span in somewhat the manner of a corkscrew.

. . . At the time it appeared that should the wind die down, the span would perhaps come to rest. . . . I [then] called the Weather Bureau . . . and was informed that the barometer was rising and in all probability the wind would quiet later in the day. . . .

I was then informed that a panel of laterals in the center of the span had dropped out and a section of concrete slab had fallen. I immediately went to the south side of the plaza. . . . The bridge was still rolling badly about the center as it had been doing previously. I returned to the toll plaza and from there observed the first sections towards each tower rapidly fall out. . . . Shortly thereafter . . . coinciding with the dropping of the sections of the center span, I observed the side span settle rapidly and was momentarily expecting the towers to come down. I did not observe the exact time that the center fell out although I was later informed that it was 11:10.[8]

In the wake of the Tacoma Narrows calamity, the U.S. Congress established a three-man commission appointed through the Federal Works Agency to investigate the failure. Ammann was asked to head the team, which included Theodore von Karman of the California Institute of Technology and Glenn B. Woodruff, who had served as the chief design engineer of the Golden Gate Bridge. Leon S. Moisseiff, designer of the Manhattan Bridge and a leading figure in deflection theory, had designed the failed bridge.

The commission's exhaustive investigation of the Tacoma Narrows Bridge's behavior included laboratory tests, site investigations, and statistical analysis. In addition, wind tunnel tests were conducted—a laboratory technique that at the time was barely a decade old. The commission's findings, published in March 1941, set off a shock wave in the engineering community. Ammann and his team concluded that Moisseiff's application of the deflection theory had unwittingly led to a road deck configuration with aerodynamic qualities somewhat like the wing of an airplane. When a certain wind velocity moved across the suspended road deck, strong negative pressure formed on the upper side. As a result, the leeward edge of the deck pointed slightly upward and lifted vertically before the gravitational force on cables and deck returned the structure to its starting position, where it began another constrained liftoff. The commission itemized a number of factors that had contributed to the bridge's excessive instability: the deck, at just two lanes wide, was simply too narrow and too lightweight for its length; the side spans were too long in relation to the clear span; and the anchorages were too far from the towers. As a result of the commission's findings, Ammann developed a corollary to the deflection theory that provided a statistical method of predicting aerodynamic stability. It considered the ratio of center span to side span to width and the effect of weight, cable sag, and stiffening girders.

After the Tacoma Narrows report was released, engineers who had designed bridges using the deflection theory subjected them to a new battery of calculations. Ammann reviewed the design of the Bronx–Whitestone Bridge, which closely followed the design of the Tacoma Narrows Bridge and opened six months before traffic began rolling over the ill-fated bridge in Washington. Ammann found his bridge's aerodynamic stability satisfactory even though the bridge swayed noticeably under certain wind conditions. Despite his conclusion, the motoring public would no longer tolerate perceivable deck motion. Against his wishes, Ammann was forced to retrofit the Bronx–Whitestone structure with stiffening trusses in 1946.

The same year, Ammann's son Werner—one of Ammann & Combs's original partners—returned from military duty to the New York practice after serving in the South Pacific during World War II. Although he, his father, and the other four members of the original consulting group had intended to keep the office small, their work volume had steadily grown, and with it the number of people on the payroll. A redirection seemed in order: either scale back or appreciably enlarge. In 1947, Othmar H. Ammann & C. C. Combs joined forces with the Wisconsin-based firm of Charles S. Whitney to form Ammann & Whitney, Consulting Engineers. The merger created one of the nation's largest civil engineering practices. Whitney was a specialist in reinforced concrete design and had established a successful practice in Milwaukee. (Ammann had met Whitney thirty years earlier in Lindenthal's office, where both had worked on the Hell Gate Bridge.) This auspicious union brought together many of the country's best designers of structural steel with unmatched expertise in reinforced concrete.

Ammann's active involvement in the practice continued uninterrupted. Through him, Ammann & Whitney was awarded the design commissions for all the long-span bridge projects undertaken by the Port Authority of New York and New Jersey and the Triborough Bridge and Tunnel Authority during the 1950s and 1960s. This final cycle of construction began when Robert Moses hired Ammann to design the Throgs Neck Bridge, a crossing for a site near the Bronx–Whitestone Bridge to relieve congestion on the 1939 structure. In approaching the design of his first long-span bridge since the Tacoma Narrows disaster, Ammann was well aware of public disdain for deck movement. To achieve new levels of rigidity, he introduced a lightweight yet remarkably stable deck structure that included lateral and transverse trusses rigidly connected. As with the plate girder stiffeners on the George Washington and Bronx–Whitestone Bridges, the interconnected truss of the single-deck Throgs Neck Bridge maintains a shallow profile and does not block motorists' views crossing over Long Island Sound. Construction on the bridge began in 1957 and was completed in 1961.

By 1957, traffic levels on the George Washington Bridge had also reached capacity, and the Port Authority hired Ammann to engineer the planned-for second (lower) deck and oversee its installation. Construction began in September 1958 and was completed in August 1962.

Ammann was also overseeing a third construction project at the time: the Verrazano-Narrows Bridge for the Triborough Bridge and Tunnel Authority. He had first been hired to design the twelve-lane bridge in 1948, but factors surrounding the project's financing kept it on hold until the mid-1950s. Destined to recapture New York City's claim as home of the world's longest suspension span, it is Ammann's crowning achievement. The deck design he had developed for the Throgs Neck Bridge was reinvestigated and expanded for the challenge posed by the double-decked Verrazano-Narrows Bridge's 4,260-foot clear span. The result was a uniquely formed, rigidly connected three-dimensional frame of unprecedented stiffness. It was the third seminal innovation Ammann made in the ongoing development of suspension bridge design. His other two significant contributions—the world's first monolithic plate-steel towers designed for the Bronx–Whitestone Bridge, and the demonstrated reliability of the deflection theory at the George Washington Bridge—in addition to the Verrazano-Narrows's road deck system, all set new standards in structural strength and economy.

Ammann was one of the most inventive bridge designers in history. He was also one of the most ardent practitioners of aesthetics. In his words, "it is a crime to build an ugly bridge." He possessed a highly developed eye for simple yet dynamic form, as illustrated by the Verrazano-Narrows Bridge. With a suspended load 75 percent greater than that of the

1.14 Aerial photograph of New York in 1964 showing the Verrazano-Narrows Bridge at the entrance between the Upper and Lower Harbor. (Photographer unknown; courtesy of the Museum of the City of New York)

Golden Gate Bridge, its pristine profile belies the astounding complexity of the bridge's engineering. Rule of thumb dictates that complex design problems yield complex solutions. Ammann's steadfast impulse toward classical restraint conditioned his approach to deflection theory in an effort to reduce and refine this bridge's line, detail, and interplay of parts. From the speaker's dais on opening day, Mayor Robert F. Wagner described the bridge as "a structure of breathtaking beauty and super engineering."

Ammann's legacy, however, reaches well beyond the achievements of a formalist and technician. A man of unshakable determination and genuine humility, his determination drove him to excel even in the face of prolonged discouragement, and his humility dictated that the focus of his drive be something greater than himself. Had his ambition been aimed at the pursuit of money or the desire for notoriety, his life would have had a very different trajectory. Instead, he spent his energy striving to give himself to engineering, and by extension to the community at large. That is why this soft-spoken genius had no qualms spending fourteen years in public service at the very time his creative forces and physical stamina were at their peak. As a civil servant he founded the engineering departments of the Port Authority of New York and New Jersey and the Triborough Bridge and Tunnel Authority and oversaw the construction of nine bridges and two tunnels—a grueling, thankless responsibility—bringing quality and dependability to the civic realm while imposing controls that saved the public vast sums of money. He continued to serve the public with the same integrity and exacting standards as a private practitioner.

When he was nearing his eightieth birthday, Ammann began to find the daily shuttle between home in Boonton and his office in the city tiresome. To avoid the commute, he and his wife took an apartment at the top of the Carlyle Hotel in Manhattan, with breathtaking views in three directions. From his bedroom looking north, Ammann could see the George Washington Bridge stretching across the Hudson River. From the eastern-facing windows in the living room, his East River crossings—the Triborough, Bronx–Whitestone, and Throgs Neck Bridges—could be spotted, as well as the Harlem River Pedestrian Bridge (the "Little Green Bridge"). Looking to the southwest from the den, most of the Bayonne Bridge over the Kill Van Kull could be seen. The den windows also offered views of the Verrazano-Narrows Bridge, twelve miles due south, across New York Harbor. In fact, until the bridge opened on November 21, 1964, Ammann had the pleasure of following its construction from this nighttime retreat. He kept a telescope handy, much to the delight of his guests. Though he never admitted it, Ammann must have taken great satisfaction in surveying his accomplishments. His life's work was vividly reflected in six long-span highway bridges and one long-span pedestrian facility that collectively graced the urban landscape at every corner of the metropolis.

No civil engineer gave more bridges to a single city than Ammann did to New York. This achievement is all the more remarkable when the scale of the individual structures is considered. Ammann knew that his careful attention to detail and outstanding structural intuition did not altogether explain his success. Many prominent figures in engineering history had unwittingly designed bridges that failed. He told an audience of civil engineers in 1953:

The thought I should like to leave with you is that bridge engineering is not, as popularly assumed, an exact science. While ordinary structures are closely controlled by ample experience and experiments, every structure which projects into new and unexplored fields of magnitude involves new problems, for the solution of which neither they nor physical experience can furnish an adequate guide. It is then that we must rely largely on our judgement and if as a result errors or failures occur we must accept them as a price for human progress.[9]

Gay Talese, then a reporter for the *New York Times,* interviewed the engineer at his apartment in the Carlyle soon after the Verrazano-Narrows Bridge opened. The venerable and accomplished eighty-five-year-old was asked the source of his success as a design innovator. Ammann volunteered that he had been "lucky."[10] When his wife tried to gently object, Ammann intervened with quiet authority and restated: "lucky."

1.15 Othmar H. Ammann in 1961, at eighty-two.
(Photographer unknown; courtesy of the Port Authority
of New York and New Jersey)

The George Washington Bridge

The George Washington Bridge is the most significant long-span suspension bridge of the twentieth century. Nearly twice as long as any bridge built before it, the roadway carried between the towers stretches thirty-five hundred feet, the length of fifteen city blocks. To grasp the breathtaking magnitude of this engineering feat, imagine New York's Fifth Avenue between Forty-second and Fifty-seventh Streets lifted and suspended in air six stories above a river. You must also imagine Fifth Avenue to be twice its width and thronged curb-to-curb with automobiles, buses, and trucks. To engineer the unprecedented span, Ammann broke with convention and designed the structure in accordance with deflection theory. The resulting structural configuration afforded enormous cost savings while articulating a revolutionary bridge form of great elegance.

Construction crews broke ground for the George Washington Bridge on October 21, 1927, and the facility opened to traffic on October 25, 1931—eight months ahead of schedule. It was estimated that the bridge would cost $60 million to build; thanks to engineering precision and efficiencies in construction, it ended up costing a million dollars less.

When plans for the bridge were developed during the 1920s, relatively few motor vehicles traveled the highways. Even so, it was clear to transportation planners that the age of the automobile was imminent and that traffic levels would mushroom over the next thirty to forty years. The George Washington Bridge was therefore designed with a built-in capacity for expansion. In the beginning, it had six lanes for motor vehicles and two promenades shared by pedestrians and bicycles, and a thirty-two-foot strip running down the center of the deck was left unpaved. As traffic volume mounted, in 1946 the center of the deck was paved to make two additional lanes. Twelve years later, construction on the lower level began. Its six additional lanes opened in 1962, as did a spacious new bus terminal at the Manhattan interchange plaza.

The George Washington Bridge significantly facilitated regional travel, allowing vehicles moving between New Jersey and New York to connect to the Bronx, Westchester

Economics and utility are not the engineer's only concerns. He must temper his practicality with aesthetic sensitivity. His structures should please the eye. In fact, an engineer designing a bridge is justified in making a more expensive design for beauty's sake alone. After all, many people will have to look at the bridge for the rest of their lives. Few of us appreciate eyesores, even if we should save a little money by building them.
—Othmar H. Ammann, 1958

2.1 East tower of the George Washington Bridge. (Photo, Jet Lowe; courtesy of the Historic American Engineering Record, National Park Service)

County, and New England without passing through the most congested parts of Manhattan. With the opening of the Triborough Bridge in 1936, the George Washington brought Long Island into a nearly seamless network of regional highways. Not surprisingly, the George Washington is the world's busiest bridge. With its fourteen lanes, no other bridge has a greater capacity. Indeed, no bridge accommodates a higher volume of traffic on a daily basis.

During its early years of operation, the bridge was visited by Le Corbusier, a leading figure in the International school and one of the twentieth century's most celebrated architects. His emotions were deeply stirred when he first crossed the bridge in an automobile, and his recollection probably speaks for most people experiencing a trip across the top deck.

The George Washington Bridge over the Hudson River is the most beautiful bridge in the world. Made of cables and steel beams, it gleams in the sky like a reversed arch. It is blessed. It is the only seat of grace in the disordered city. It is painted an aluminum color, and between water and sky, you see nothing but the bent chord supported by two steel towers. When your car moves up the ramp the two towers rise so high that it brings

2.2 View from within the east tower. (Photo, Jet Lowe; courtesy of the Library of Congress, HAER Collection)

2.3 A view from Fort Tryon Park soon after the bridge opened to traffic in 1931. The lower deck was not constructed until 1959–1962. (Photographer unknown; courtesy of the Library of Congress)

2.4 The bridge began operation with six lanes for vehicles and two for pedestrians and bicycles. Initially left unpaved, a two-lane width down the center of the road deck was paved in 1946 to accommodate the ever increasing volume of traffic crossing the bridge. (Photo, Underwood & Underwood; courtesy of the Library of Congress)

2.5 The double-deck bridge in 1986, from a perspective similar to that in figure 2.3. The addition of the lower road deck in 1962 realized Ammann's vision for the bridge as first conceived in 1923. (Photo, Jet Lowe; courtesy of the Library of Congress, HAER Collection)

you happiness; their structure is so pure, so resolute, so regular, that here, finally, steel architecture seems to laugh. The car reaches an unexpectedly wide apron; the second tower is very far away; innumerable vertical cables, gleaming against the sky, are suspended from the magisterial curve which swings down and then up. The rose-colored towers of New York appear, a vision whose harshness is mitigated by distance.[1]

TOWER DESIGN

Ammann relished the architectural qualities of the Brooklyn Bridge, and he particularly admired the power of its massive granite towers. Although he longed to design all-masonry towers for the George Washington Bridge, he readily conceded that they were no longer feasible for long-span bridges: they would weigh too much and cost too much.

The engineer found himself torn between the technological imperative of his day — which demanded lightweight steel towers — and a personal and altogether artistic impulse to use stone. In his struggle to reconcile these opposing influences, Ammann developed plans for a hybrid form of tower that used steel as the primary material and concrete as a secondary one. The plan called for hundreds of steel "twigs" pulled apart and braced together to form an armature that approximated the volumetric profile of a traditional stone tower with two portal arches, one above and one below the road deck. The steel would support the bridge during construction; a concrete membrane poured at the surface of the steel network would provide added strength to support the weight of traffic after the facility opened, and it would furnish the masonry surface Ammann desired. It seemed a perfect marriage between modern materials and a romantic vision.

As engineering details were developed, it became clear that the concrete would be structurally superfluous: the steel frame alone would be sufficiently strong to carry both the tower's dead and live loads. With that realization, Ammann scrapped plans for a composite structure and asked his consulting architect, Cass Gilbert, to prepare plans for an ornamental stone facade that would simply be bolted to the outer face of the steel frame. After they had worked out the aesthetic and technical challenges of the facade, however, Ammann was foiled again. This time, economics undermined the masonry: a dramatic downturn in the region's economy brought on by the Great Depression caused the project's governing board to retract its earlier authorization and deny funding for the stonework. With the steel up, it was simply too late for Ammann to again rethink the tower design.

Measuring 604 feet from the water level, the bridge's towers were taller than most sky-

2.7 Iron workers connecting the prefabricated steel members of the bridge's east tower. (Photographer unknown; courtesy of the Port Authority of New York and New Jersey)

2.6 Construction of the towers. In the foreground, the Manhattan anchorage is under construction, with its eyebar beams still exposed. (Photographer unknown; courtesy of the Port Authority of New York and New Jersey)

2.8 Detail of steel connections. (Photographer unknown; courtesy of the Port Authority of New York and New Jersey)

2.9 Tower detail at the springing of the upper portal arch. (Photo, Jet Lowe; courtesy of the Library of Congress, HAER Collection)

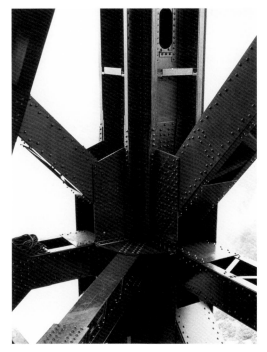

scrapers of their day. Twenty thousand tons of steel were used in each. The steel sections of the towers were manufactured off site, transported by rail, towed by barge, and hoisted into position with cranes and mechanical lifts designed especially for the job. Sections weighed between 37 and 80 tons and were attached with rivets. (There are 475,000 rivets in each tower, accounting for 325 tons of deadweight.)

The frustration of his efforts with the towers radically changed Ammann's sensibilities. Although the Brooklyn Bridge inspired him throughout his life, there is no evidence that he ever again so much as sketched a rough design for a masonry tower. Instead he introduced technical and aesthetic innovations in the design of slender steel towers, demonstrating how passionately he came to favor steel over masonry in his own work.

A THWARTED PLAN FOR STONEWORK

The George Washington Bridge is one of America's most conspicuous unfinished works of architectural engineering. Designed to be covered with granite, the closely knit network of steel making up the towers was meant to provide the armature on which would hang thousands of pieces of dimensional stone.

Ammann chose stone to strike a harmonious balance between the architecture of his structure and the natural grandeur of the site. The bridge meets the New Jersey coast above the chiseled cliffs of the Palisades. On the New York shore, dramatic granite outcroppings carve their way from water's edge to high plateau. Ammann's towers were conceived as natural extensions of this rugged landscape. While formally related to the opposing cliffs, the stone towers promised to pose a perfect foil to the shiny steel of the cables and road deck hovering over the water.

The governing board responsible for approving all matters concerning the bridge project accepted preliminary plans for steel towers hung with stone. Consequently, the office of Cass Gilbert pushed forward under Ammann's direction and prepared numerous schemes that studied varying approaches to massing and stylizing the sheathing. In the beginning of their studies, historical models were favored: classical, baroque, and Gothic. Design trends, however, were shifting in America. At the same time the bridge was on the drawing board, modernism was beginning to supersede revivalism in popularity. The final design selected for the stonework reveals a stripped-down, simplified profile characteristic of the Art Deco style (fig. 2.14).

Ammann and Gilbert objected passionately to the governing board's cancellation of the

2.10 Steel construction on the west tower. (Photographer unknown; courtesy of the Port Authority of New York and New Jersey)

2.11 Rendering of a design scheme for cladding the bridge's towers with stone. Note the formal esplanade, garden, and yacht basin also under consideration. (Rendered by F. G. Stickel, Office of Cass Gilbert, Architect; courtesy of the Museum of the City of New York)

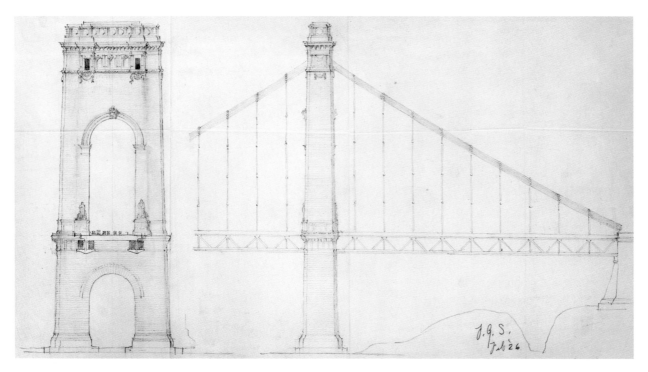

2.12 Baroque style tower scheme. (Rendered by F. G. Stickel, Office of Cass Gilbert, Architect; courtesy of the New-York Historical Society)

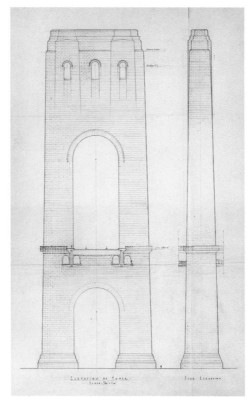

2.14 Art Deco style tower scheme. (Delineated by the Office of Cass Gilbert, Architect; courtesy of the New-York Historical Society)

2.13 Gothic style tower scheme. (Rendered by F. G. Stickel, Office of Cass Gilbert, Architect; courtesy of the New-York Historical Society)

2.15 Rendering of a scheme for an exposed steel structure. (Delineated by the Office of Cass Gilbert, Architect; from the Linda Hall Library, courtesy of the American Society of Civil Engineers)

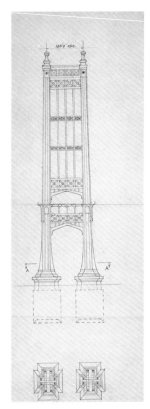

left

2.16 Front elevation of the exposed steel tower scheme. (Delineated by the Office of Cass Gilbert, Architect; courtesy of the New-York Historical Society)

right

2.17 A steel structure designed to be concealed remains exposed, an enduring testament to the public's embrace of the Machine Age aesthetic. (Photo, Jet Lowe; courtesy of the Library of Congress, HAER Collection)

ornamental stone as a cost-saving measure. After offering arguments in their final pleas that even they must have considered farfetched, once it became clear that stone would not be provided for the towers neither engineer nor architect publicly expressed regret. Happily, the machined profile and transparency of the unfinished construction was quickly and enthusiastically embraced by the public.

Earlier, however, while assiduously investigating patterns for decorative stonework, Ammann and Gilbert had also looked at several designs for exposed steel towers. Had they known that the project was destined to have steel towers, there is little doubt that they would have configured them in a slender shape rendered with a few heavy lines, similar to the designs they considered (figs. 2.15, 2.16).

CABLE SPINNING

The bridge's engineering team designed two suspension systems that were deemed structurally equivalent. One system called for an eyebar-type network. Also known as suspension chains, these are made by bolting together hundreds of thin metal bars at eyelets on each end. Early renderings for the proposed George Washington Bridge typically depicted a structure with suspension chains (fig. 2.11).

A spun-cable suspension system provided an alternative to chains. With a spun cable, small individual wires are spun in air from anchorage to anchorage and over the towers by means of a traveling wheel. Once the requisite number of wires are accumulated, they are bundled into strands that are in turn compacted and tightly wrapped with wire to form the completed cable.

Because Ammann regarded suspension chains and cables as equally effective, contractors were encouraged to offer bids for either. The contract was awarded to the lowest bidder, so the choice of suspension system was determined in the marketplace. The low bidder and contract recipient was a cable manufacturing and spinning operation: John A. Roebling & Sons, a firm founded by the designer of the Brooklyn Bridge. Its New Jersey plant manufactured the 107,000 miles of steel wire used to make the George Washington Bridge's main suspension cables. Roebling & Sons iron workers then spun the bridge's two pairs of three-foot-diameter cables. Each cable is made of 61 large strands, and each of these strands is spun from 434 wires wound together over the river. The temporary footbridges (catwalks) seen in construction photographs were built about four feet below the main cable. Supported by the first spun wires, this bridge-within-a-bridge was used by con-

above left

2.18 Temporary walkway construction as it began from the New Jersey side of the suspension span. (Photographer unknown; courtesy of the Port Authority of New York and New Jersey)

above right

2.19 Threading wire around the eyebars of the anchorage system. The spinning wheels in the photograph were used to carry individual strands of wire from anchorage to anchorage, building up the two pairs of three-foot diameter suspension cables from which the road deck is hung. (Photographer unknown; courtesy of the Port Authority of New York and New Jersey)

right

2.20 Carriage for handling compactors, cable bands, and suspenders. (Photographer unknown; courtesy of the Port Authority of New York and New Jersey)

far right

2.21 Bundling the wires that wrap around one eyebar anchor. (Photographer unknown; courtesy of the Port Authority of New York and New Jersey)

struction workers while spinning the cable and attaching the suspender ropes that hang down to support the road deck.

Spinning the first strand of cable for a long-span suspension bridge tends to draw spectators. To begin the Niagara Suspension Bridge, Charles Ellet, Jr., its first engineer, sponsored a kite flying contest, offering five dollars to the first American child who could fly a kite to the Canadian bank of the river. Homer Walsh successfully managed the feat across

2.22 Aerial photograph taken during the cable-spinning phase of construction. (Photographer unknown; courtesy of the Port Authority of New York and New Jersey)

2.23 Installing a temporary walkway for cable spinning. (Photographer unknown; courtesy of the Port Authority of New York and New Jersey)

2.24 Suspender cables ready for road deck attachment as viewed from within the anchorage. (Photographer unknown; courtesy of the Port Authority of New York and New Jersey)

2.25 Detail of tower, suspension cables, and suspender cables. (Photo, Dave Frieder; courtesy of Dave Frieder)

2.26 View from the pedestrian walkway where the suspension cables dip to the bottom of their curve. (Photo, Jet Lowe; courtesy of the Library of Congress, HAER Collection)

the 759-foot gorge that separated the towers of the embryonic bridge. Thanks to the young man's successful aviation, a team on either side of the river had hold of a kite string. A somewhat sturdier rope was tied to the string; the kite string was then pulled back across so that the two teams were now connected with a rope. The process continued with ever sturdier ropes until the first wire was pulled across the chasm. The wire was then hoisted over the towers and tethered to the anchorages. Before additional wire was spun across the channel with the flywheel, the engineer amazed the public with a heart-stopping demonstration of faith: hanging from a basket attached by a pulley to the single wire, Ellet propelled himself from one side to the other. He proved the wire's strength, thrilling an adoring crowd assembled on the banks of the gorge.

Cable spinning commenced less playfully at the George Washington Bridge. River traffic on the Hudson was closed for part of a day as barges loaded with spools of cable laid wire along the bed of the river from tower to tower. The starter cables were then mechanically hoisted to the cable saddles at the tops of the towers and attached to the eyebar an-

2.27 Looking between a pair of suspension cables to the pedestrian walkway. Above the main cables are safety wires used by inspectors and maintenance workers. (Photo, Jet Lowe; courtesy of the Library of Congress, HAER Collection)

chors of the anchorages. Once the first wires were in position, the traveling wheels went to work.

From its simple beginnings with the Niagara Suspension Bridge, which opened in 1855, spinning efficiency improved greatly. The 3,600 tons of cable wire for the Brooklyn Bridge (opened in 1883) took twenty-one months to spin; the 6,400 tons for the Manhattan Bridge (1909), four months; and the astonishing 28,100 tons for the George Washington Bridge (1931), a mere ten months.

WHERE CABLES DISAPPEAR FROM SIGHT

Because the anchorages at the George Washington are beneath the level of the road, the main suspension cables are threaded through the deck on their way to being anchored. It's both astonishing and curious to see the powerful cables simply disappear from sight. It was not, however, the intention of the project's consulting architect that the intersection of the cables and the road be left unarticulated. Indeed, Gilbert and his design team felt challenged to provide a pleasing and expressive architectural detail at these points. Between 1925 and 1926, Gilbert's design office prepared numerous preliminary studies for stylizing the cable ends. The general effect intended by the architectural treatment was twofold. First, a conspicuous sculptural form was prescribed because it would bring the cables to a deliberate point of termination before they disappeared from sight—metaphorically, the period at the end of a sentence. Second, the imagery and formal composition of the mass was meant to convey the dynamic tensile forces at work in the cables as they flex to support the weight of the road deck—an allegorical device to convey a genuine physical state.

The accompanying selection of cable-end details represents the range of whimsical forms considered by the designers. From the "Winged Tire," directed to resist the cables' pull, to the "Laboring Group" frozen in its struggle to hold back the cables in a never ending game of tug-of-war, all the proposed schemes were placed on long, low pedestals intended to soften the geometric transition between diagonal cable and horizontal road deck.

The stylized cable ends met the same fate as nearly all the bridge's ornamental detailing: they were canceled due to unforeseen economic constraints. With a bow to pragmatism, a simplified steel fitting eases the juncture of cable and road surface. Today, a high fence surrounds the cable ends to prevent would-be daredevils from scaling the bridge.

Detailing where cables visibly connect with other bridge components would remain a formal preoccupation with Ammann and his consulting architects in succeeding suspen-

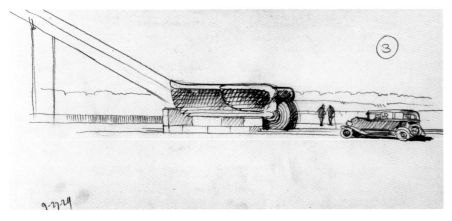

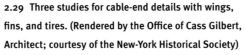

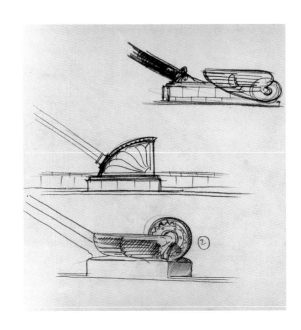

2.28 "Winged Tire" scheme for cable end. (Rendered by the Office of Cass Gilbert, Architect; courtesy of the New-York Historical Society)

2.29 Three studies for cable-end details with wings, fins, and tires. (Rendered by the Office of Cass Gilbert, Architect; courtesy of the New-York Historical Society)

top

2.30 Griffin scheme for cable end. (Rendered by John T. Cronin, Office of Cass Gilbert, Architect; courtesy of the New-York Historical Society)

bottom

2.31 Scheme for the cable end incorporating a replica of an Assyrian Colossus. (Rendered by the Office of Cass Gilbert, Architect; courtesy of the New-York Historical Society)

sion bridge designs. In a subsequent project, the Triborough Suspension Bridge (1936), architect Aymar Embury II shifted away from Gilbert's decorative tendencies toward design solutions that were ever more prosaic yet more refined. Ammann's ongoing search for a more essential resolution culminated in his last bridge, the Verrazano-Narrows. Here, the Euclidean geometry of the anchorage echoes the inclination of the cable, and rather than disappearing into the road, cables confidently connect to the anchorage end-walls (fig. 7.12). All construction elements are powerfully resolved with an absolute economy of means.

GAMBLING WITH AN UNTESTED THEORY

The suspended road deck is the most vulnerable element of a suspension bridge. Traffic and wind movement tend to cause decks to sway from side to side and heave up and down. If movement becomes too great, the deck will collapse. To counter this tendency, early-nineteenth-century engineers stabilized their structures with guy wires that ran from the towers to the road deck or from the deck to the ground. (Roebling's Niagara Railroad Bridge had guy wires anchored to both the towers and the ground.) Later in the century bridge designers began attaching deep trusses to the sides of their suspended road decks for stabilization. As suspension bridges got longer they also got heavier. The largely untested

2.33 Profile of the single-level road deck taken several years after the bridge opened to traffic. (Photo, Irving Underhill; courtesy of the Library of Congress)

2.34 Assembly of the plate-girder stiffening trusses beneath the road deck. (Photographer unknown; courtesy of the Port Authority of New York and New Jersey)

"deflection theory," circulating at the time Ammann designed the George Washington
Bridge, held that as the weight per linear foot of long-span bridges increased, the need for
stiffening decreased because the greater deadweight of the structure itself would play a major role in resisting movement.

Leon S. Moisseiff was the first engineer to apply the theory to suspension bridges, using it in the design of the Manhattan Bridge, which opened in 1909. Deflection theory al-

2.36 Profile of the road deck after the completion of the lower level. (Photo, Jack Boucher; courtesy of the Library of Congress, HAER Collection)

2.37 Scaled drawing of the George Washington Bridge and surrounding land form. (Delineated by John T. Cronin, Office of Cass Gilbert, Architect; courtesy of the New-York Historical Society)

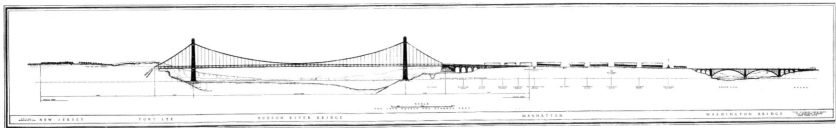

lowed him to significantly reduce the size of the bridge's structural members, thereby eliminating tons of steel and resulting in enormous cost savings. Though the Manhattan Bridge was a success, Ammann was genuinely gambling when he applied deflection theory in the design of the George Washington Bridge, with a span more than twice as long as Moisseiff's bridge. The gamble, however, seemed justified when balanced against the economic benefits. Also a success, the George Washington Bridge conclusively proved the reliability of deflection theory for suspension bridges approaching or exceeding thirty-five hundred feet in length.

The George Washington commission also afforded Ammann the opportunity to introduce a previously untested road deck assembly. Some type of stiffening device was needed for the deck even though deflection theory allowed an overall reduction in material. Moisseiff incorporated tried-and-true stiffening trusses in the Manhattan Bridge. Ammann instead used a pair of plate girders integral to the underside of the roadway—an engineering first. The girders contribute to the graceful appearance of the bridge while affording travelers unobstructed views from the bridge, which most stiffening-truss configurations don't allow.

From a distance, the single-deck span seems slight, even frail. With a deck only twelve feet deep—and a depth-to-span ratio of 1:120—it appears as if the road deck could fly away. But neither traffic patterns nor wind caused the structure to flutter. When the lower deck was added in 1962, the span's ability to resist torsional movement was further enhanced. Like the upper deck, the lower one is supported by a girder truss. The two decks are connected with an open-web truss that ties the configuration together, making it nearly as rigid as a rectangular tube.

THE NEW YORK PLAZA

The New York bridge approach was originally conceived as a grand public plaza (fig. 2.38). Traffic was to enter from several ramps leading to a spectacular central fountain flanked by beguiling statuary. Motor vehicles would circle the gushing fountain en route to the bridge or radiate from the plaza center to adjacent streets by way of ramps. Though it may have looked charming on paper, the crisscross movements involved in such a traffic arrangement would have resulted in congestion, delays, and accidents. Gilbert's original scheme, which owed much to the City Beautiful movement, now seems quaint and unrealistic in light of the demands of high-speed, high-density traffic.

The final design of the New York interchange in no way resembles the earlier proposal. As built, it separates traffic flowing in opposite directions and slow from relatively fast-moving traffic. Separate ramps carry vehicles to and from the various connecting highways, avoiding left turns and grade crossings throughout. The result is a complicated arrangement of spiraling ramps, underpasses, and overhead crossings sometimes three stories

right

2.38 Study for a monumental plaza at the approach to the east tower. (Rendered by F. Radberg, Office of Cass Gilbert, Architect; courtesy of the New-York Historical Society)

below left

2.39 Phase-one demolition for the Manhattan plaza. (Photographer unknown; courtesy of the Port Authority of New York and New Jersey)

below right

2.40 Phase-two demolition for the Manhattan plaza. (Photographer unknown; courtesy of the Port Authority of New York and New Jersey)

2.41 East tower and approach plaza. (Photo, Jack Boucher; courtesy of the Library of Congress, HAER Collection)

2.42 Study for the bridge's side-span statuary and stone tower. (Rendered by John T. Cronin, Office of Cass Gilbert, Architect; courtesy of the New-York Historical Society)

left

2.43 East tower and approach plaza. (Photo, Jet Lowe; courtesy of the Historic American Engineering Record, National Park Service)

right

2.44 The Manhattan street grid is largely uninterrupted by the world's busiest bridge. (Photo, Jet Lowe; courtesy of the Historic American Engineering Record, National Park Service)

high. It more closely resembles an efficient machine than an urban plaza scaled to the speed of a horse-drawn carriage.

Construction of the Manhattan interchange entailed carving into Fort Washington, a fully developed neighborhood. Although it was inevitable that this mammoth bridge would clash with the pattern of a nineteenth-century residential district—indeed, it altered it irrevocably—it actually gobbled up little property that was useful for building. Furthermore, the impact on local traffic remains remarkably slight. The main flow to and from the bridge does not touch the neighborhood streets, instead being spun out on ramps to the parkways along the Hudson River or underground via Interstate Route 95.

THE NEW JERSEY PLAZA

Conditions on the New Jersey side of the bridge were considerably different. Fort Washington was a densely populated neighborhood; Fort Lee, New Jersey, was largely un-developed (fig. 2.22). With no existing structures posing a constraint, bridge planners were free to configure the plaza to their ideal standards for safety and efficiency.

2.45 River Road leading to the New Jersey approach. (Photographer unknown; courtesy of the Port Authority of New York and New Jersey)

2.46 Study for New Jersey plaza showing toll booths and maintenance building. (Rendered by John T. Cronin, Office of Cass Gilbert, Architect; courtesy of the New-York Historical Society)

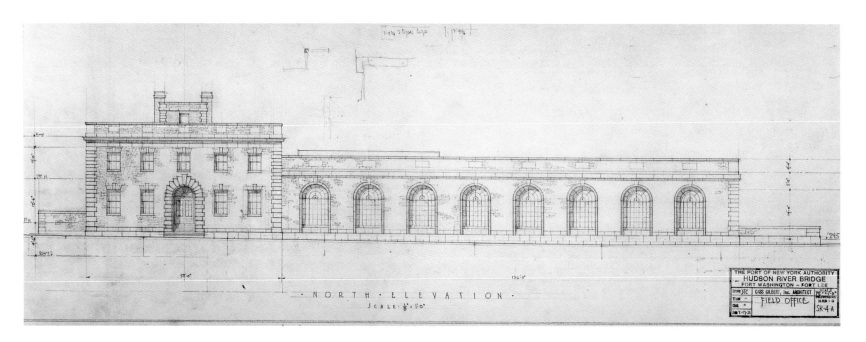

N · O R T H · E L E V A T I O N ·

SCALE: ⅛" = 1'-0"

THE PORT OF NEW YORK AUTHORITY
HUDSON RIVER BRIDGE
FORT WASHINGTON — FORT LEE
CASS GILBERT, Inc. ARCHITECT
FIELD OFFICE
SK-4-A

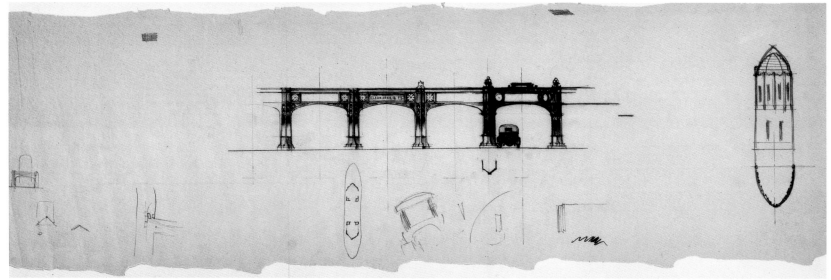

top

2.47 **Study for a maintenance facility designed in the classical style.**

(Delineated by John T. Cronin, Office of Cass Gilbert, Architect; courtesy of the New-York Historical Society)

bottom

2.48 **Sketch study for toll facility. (Rendered by Cass Gilbert; courtesy of the New-York Historical Society)**

2.49 Photograph of the New Jersey plaza taken soon after the bridge opened. (Photographer unknown; courtesy of the Port Authority of New York and New Jersey)

2.50 Detail of original toll booths. (Photographer unknown; courtesy of the Port Authority of New York and New Jersey)

2.51 The New Jersey approach and toll plaza. (Photo, Jet Lowe; courtesy of the Historic American Engineering Record, National Park Service)

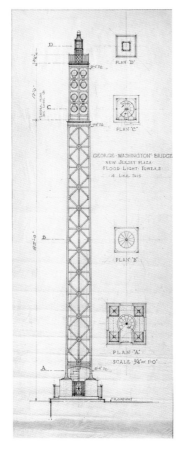

2.52 View from the west tower to the New Jersey approach plaza. (Photo, Jet Lowe; courtesy of the Library of Congress, HAER Collection)

2.53 Study for a floodlight tower. (Delineated by the Office of Cass Gilbert, Architect; courtesy of the New-York Historical Society)

The New Jersey approach incorporates the bridge's only toll plaza. The architectural design of collection booths, floodlights, and maintenance facilities was entrusted to Gilbert's office. At the beginning of the design process, the architects advocated building materials and stylistic forms that recalled grand civic architecture from the past (fig. 2.47); but as time went on they came to appreciate the uncontrived form of the bridge itself. Toll booths, floodlights, and support buildings took on a clean, machined appearance, harmonizing with the masonry foundations and exposed steel of the bridge (figs. 2.46, 2.48).

Gilbert's light towers for the toll plaza, with their open framing threaded with a spiral stair, sounded a playful note amid the bridge's overwhelming grandeur (fig. 2.53). Unfortunately, none of Gilbert's work at the George Washington toll plaza remains, having been lost in the facility's expansion. He designed a similar series of floodlight towers for Ammann's Lincoln Tunnel of 1934, however, and these towers are still in service and are well maintained.

A PARKWAY APPROACH

It so happened that the bridge's siting placed it in Manhattan's Fort Washington Park, the northern continuation of Riverside Park. Beginning on West Seventy-second Street, Riverside Park's winding paths, rustic retaining walls, and informally arranged plantings follow the English pastoral style. Frederick Law Olmsted, the principal landscape architect for Central Park, was a guiding force in the design of both Riverside Park, which ends at 158th Street, and Fort Washington Park, which adjoins it and continues the scenic public parkland to the base of the George Washington Bridge. Today, both Riverside Drive and the Henry Hudson Parkway (often mistakenly called the West Side Highway) course their way through the parks.

Not surprisingly, Ammann and his design team regarded the Manhattan approach from the south as the most architecturally splendid, seeing the bridge as the focal point for all northbound traffic. Working with previously established landscape features, they sought to enhance the carefully composed visual corridor with crossover bridges, vehicular ramps, esplanades, tunnels, and retaining walls that were sympathetically detailed to harmonize with the surroundings. Of the many new roadway features designed in conjunction with the bridge, none was more painstakingly considered than the tunnel configuration at the front of the New York anchorage through which Riverside Drive passes. Both the exterior and the interior of the underpass were carefully studied, as were the number and arrangement

top
2.54 Riverside Park joins Fort Washington Park at the foot of the east tower.
(Photo, Jet Lowe; courtesy of the Historic American Engineering Record, National
Park Service)

bottom
2.56 Aerial view of Fort Washington Park in the foreground, looking across the
river to the New Jersey side. (Photo, Jet Lowe; courtesy of the Historic American
Engineering Record, National Park Service)

top
2.55 Approaching the George Washington Bridge along Riverside Drive in 1938.
(Photo, Irving Underhill; courtesy of the Library of Congress)

bottom
2.57 Study for the landscape design at Fort Washington Park.
(Rendered by John T. Cronin, Office of Cass Gilbert, Architect; courtesy of the
New-York Historical Society)

top left

2.58 Study for a crossover bridge above Riverside Drive. (Rendered by John T. Cronin, Office of Cass Gilbert, Architect; courtesy of the New-York Historical Society)

top right

2.59 Study for a retaining wall and stairs adjacent to the Manhattan anchorage. (Rendered by John T. Cronin, Office of Cass Gilbert, Architect; courtesy of the New-York Historical Society)

2.60 Study for the east tower's underpass. (Rendered by John T. Cronin, Office of Cass Gilbert, Architect; courtesy of the New-York Historical Society)

2.61 The George Washington Bridge. (Photo, Jet Lowe; courtesy of the Library of Congress, HAER Collection)

of the tunnel openings. As built, a single-arched opening faced with brick gives passage to the parkway.

New leisure and recreational facilities, such as esplanades and a yacht basin, were planned for the base of the east tower. As with so many of the civic amenities designed by Gilbert's team, the esplanades and yacht basin have yet to find a life beyond the drawing board.

The Bayonne Bridge

3

The city of Bayonne, New Jersey, is separated from Port Richmond, New York, by a shipping channel that early Dutch settlers named the Kill van Kull. Running between the southern tip of the Bayonne Peninsula and the northern edge of Staten Island, the channel joins the upper bay of New York Harbor with Newark Bay, which in turn connects to the Passaic and Hackensack Rivers to the north and the Arthur Kill to the south. The Kill van Kull is one of the busiest commercial waterways in the world; the tonnage shipped through it exceeds that of the Suez Canal.

Commercial ferries began operations across the channel in the 1820s. Over time, towns grew on either end of the line, developing a system of streets, rail lines, and transportation terminals as extensions of the ferry service. When a bridge was planned that would supersede the ferries, it was only natural for its crossing to overlay the ferry route. Bridge planners explained that following the lines of the ferries involved the destruction of the least amount of taxable property and was most convenient to major street connections on both sides of the waterway.

Opening to traffic in November 1931, the Bayonne Bridge over the Kill van Kull was the last of three bridges built by the newly formed Port of New York Authority as highway connections between the New Jersey mainland and the borough of Staten Island in New York. The other two bridges, designed by Waddell and Hardesty, Consulting Engineers, were the Outerbridge Crossing Bridge (1928), the southernmost connection, and the Goethals Bridge (1928), which crosses roughly midpoint on the western shore of Staten Island. The Bayonne Bridge provides a link at the northern end of the island. Together, the three bridges carry fourteen lanes of traffic.

All three were planned as part of the Port Authority's emerging vision for a unified transportation network that would knit together a variety of travel modes in the greater metropolitan area. The primary ambition for these crossings was to forge a connection between New Jersey and Long Island, with Staten Island as a stepping-stone in the circum-

The Port Authority recognized the fact that its structures must not only be useful, but they must also conform to the aesthetic sense. This was one of the motives for the selection of an arch spanning the entire river in one sweeping graceful curve.
—Othmar H. Ammann, 1931

3.2 Aerial view of the bridge with Staten Island in the foreground and New Jersey in the background. (Photo, Jet Lowe; courtesy of the Historic American Engineering Record, National Park Service)

ferential loop. Although Staten Island residents and workers would account for some of the bridge traffic, most travelers would simply pass through en route to the region's airports, harbors, rail lines, and north–south and east–west highway corridors.

The Port Authority—whose jurisdiction is confined to interstate transportation—built the western links for a circumferential loop by connecting New Jersey to Staten Island with its three bridges, but the authority could only presume the speedy construction of an eastern crossing linking the intra-urban boroughs of Staten Island and Brooklyn. It was reasonable to count on a Brooklyn connection. Construction of a tunnel between Staten Island and Brooklyn under the Verrazano Narrows began in 1921. The project was soon abandoned, but the bullish planners of the Port Authority were confident that work on the tunnel would quickly resume after the completion of their interstate bridges. To their chagrin, more than thirty years passed before the connection was made with the Verrazano-Narrows Bridge in 1964. Until then, all three of the New Jersey to Staten Island toll bridges were underused and failed to generate enough revenue to be self-liquidating and self-maintaining.

opposite

3.1 Detail of the Bayonne Bridge's arch. (Photographer unknown; courtesy of the New York Public Library)

Of the three, the Bayonne Bridge has always been the most underused. The other two are more heavily traveled because they provide convenient links to the region's interstate highways (plans are currently under way for an additional six-lane bridge near the site of the Goethals). Transportation planners had a reasonable expectation of high levels of local traffic on the Bayonne Bridge, because it is convenient to the Holland Tunnel, built in 1928. They assumed that large tracts of rural Staten Island would be developed into residential communities for commuters who would use the bridge and tunnel to reach jobs in Manhattan by automobile, but a significant commuting pattern across the Bayonne Bridge and the Holland Tunnel never emerged.

So great was the optimism of planners toward the bridge's potential that it was designed for immediate expansion. Planners assumed that surging traffic levels would render its four-lane roadway inadequate within twenty-five years of its opening. The Port Authority therefore insisted that the approach viaducts and the suspended roadway be designed so that an additional three to four lanes could be surfaced on the road deck without modification to the primary structure. Furthermore, although the planning commission originally determined that accommodations for commuter rail transit should not be incorporated into the growth plan—bus transportation was expected to meet new commuter demand—powerful interests in the rail industry interceded. Capitulating to the rail industry's lobbyists, the Bayonne Bridge Commission demanded further design modifications to strengthen the structure so that two lines of light rail might be added if dictated by consumer demand. This final change to the structure's design added appreciably to its initial cost.

At the end of the twentieth century, though, the bridge was still able to accommodate the smooth flow of traffic with its original four vehicular lanes, and its capacity for expansion remains untapped.

DETERMINING FORM

Also known as the Kill van Kull Bridge, this elegant structure is the only one of Ammann's six major works that is not a suspension bridge. Sixty-six hundred feet of approach viaduct lead to an arch over the Kill van Kull; together, the viaducts and arch carry the roadway for a half mile.

Before the choice of an arch design was finalized, a suspension-type bridge was considered, as was a cantilever-type bridge. Estimates quickly revealed that a cantilever bridge

3.3 Profile of the Bayonne arch and approach viaducts. (Photo, Jet Lowe; courtesy of the Historic American Engineering Record, National Park Service)

would be far too expensive and too unattractive. Engineering and architectural studies were made for three suspension bridge configurations, all with roadways stiffened by a commonly used truss configuration known as a Warren truss. Cost estimates for a suspension bridge were comparable to those for an arch until the designer was asked to modify the proposals to accommodate the addition of two lines of rail. Stiffening the suspension bridge for rail service proved significantly more expensive than modifying the arch. Economic imperatives aside, Ammann all along had considered an arch an aesthetically superior form for the low-lying industrial landscape it would visually dominate.

The site geography is particularly well suited to support abutment foundations for an arch bridge. Geologically, Staten Island is part of the continental mainland and a continuation of the Jersey Palisades. Solid rock extends under the city of Bayonne and across the Kill van Kull to Port Richmond; lying about ten feet below sea level on the north side and twenty-five to thirty-five feet below on the south, the rock receives the thrust of the arch.

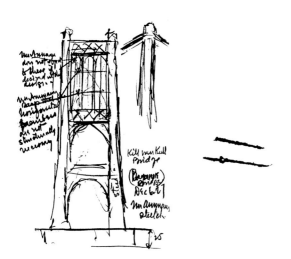

3.4 Study sketch for the steel towers of a suspension bridge scheme. (Drawing by Cass Gilbert; courtesy of the New-York Historical Society)

3.5 Study sketch for a design incorporating diagonal bracing for the steel towers of a suspension bridge scheme. (Drawing by Cass Gilbert; courtesy of the New-York Historical Society)

3.6 Study sketch of a partial profile for a suspension bridge. (Drawing by Cass Gilbert; courtesy of the New-York Historical Society)

The final structural form of the bridge (technically, a spandrel-braced arch) owes much to Lindenthal's Hell Gate Bridge. Also a trussed arch, Ammann's design accepted the theoretical premise, structural configuration, and construction methods established by the Hell Gate. The differences between the two colossal structures reflect the greater length of the Bayonne Bridge (seven hundred feet longer) and the enormous dead- and live-load capacity of the Hell Gate, which was designed to carry heavy freight trains.

THE STEEL ARCH AND SUSPENDED DECK

When it opened in 1931, with a 1,675 foot clear span, the Bayonne Bridge was the longest arch bridge in the world. Its closest rival, John Bradfield's Sydney Harbor Bridge (1932) in Australia, was under construction at the same time. (Although Sydney Harbor's has the world's broadest roadway—160 feet—with four rail lines, eight vehicular lanes, and two pedestrian walkways, Bayonne's bridge is 25 feet longer.) In 1977, the Bayonne arch lost its status as the world's longest to the New River Gorge Bridge at Fayetteville, West Virginia, with an arch 25 feet longer. The pedigree of all these bridges can be traced to the Hell Gate Bridge, Lindenthal's model of efficiency and beauty in long-span steel arch design.

The Bayonne's spandrel-braced arch comprises two trussed arches that run parallel to the outside of the suspended roadway. They are tied together across the width of the roadway with spandrel braces. The composite form is shallow at the top (thirty-seven and a half

3.7 Construction of the steel arch from the Port Richmond abutment on Staten Island. (Photo, Hoyt; courtesy of the Port Authority of New York and New Jersey)

3.8 The arch was built asymmetrically, closing beneath the peak of the arch, nearer the Staten Island shore. (Photographer unknown; courtesy of the Port Authority of New York and New Jersey)

3.9 The form of the arch is long and shallow. The truss is deep at its springing and gradually tapers toward the peak. (Photo, Jet Lowe; courtesy of the Historic American Engineering Record, National Park Service)

3.10 Attaching the road deck structure to the arch with suspender cables. (Photographer unknown; courtesy of the Port Authority of New York and New Jersey)

3.11 The road deck was erected in segments, beginning at midspan and proceeding toward the viaducts. (Photographer unknown; courtesy of the Port Authority of New York and New Jersey)

feet) and grows deeper toward the abutment (sixty-seven and a half feet); "The general outline of the arch with height decreasing from the center toward the ends was preserved principally for its pleasing appearance," according to Ammann.[1] The bottom two chords of the truss—its principal structural members—form a perfect parabolic arch. The upper arch chord and bracing act principally as stiffeners. A cut through the bottom and top chords would reveal hollow box construction, fabricated with plate steel. The bottom chord has a deeper cross section than the top and is made with manganese steel—a stronger and lighter metal than the silicon steel used in the top chord. All the weight and stresses accumulated by the arch are gathered into the bottom chord and transferred to concrete abutments. The connection between the arch and the abutment is a hinge joint made of two fifty-ton castings and a sixteen-inch-diameter steel pin; the trussed arch moves around the pins as temperature and gravity loads cause it to flex up and down.

Raising the arch was a remarkable feat of speed, efficiency, and deft execution. Ideally, building a freestanding arch is done from scaffolding. Because of the Kill van Kull's importance as a shipping lane and a defensive naval route, however, the channel could not be blocked at any time during construction, so the builders combined cantilever and temporary falsework methods.

Forty truss segments fabricated off site compose the arch. The builders put up the segments asymmetrically: first, they raised fourteen segments from the Staten Island abutment; once complete, twenty-six segments were assembled from the New Jersey side. They then closed the two segments directly above the main channel, not at the peak of the arch (fig. 3.8). During construction, each truss segment was lifted from a barge and positioned for connection by a crane with a seventy-foot boom that traveled on a track laid along the upper chord of the completed arch assembly. Each newly attached segment was cantilevered from the previous one, which was in turn steadied from beneath by a movable hydraulic jack. The arch was finally closed using the jack for alignment. The accuracy in design, fabrication, and construction was such that the arch closed with a difference of only one-half inch from its theoretical length.

The suspended deck, a network of perpendicular beams, girders, and stringers, is hung from the arch with wire rope. Sections of the primary deck structure were assembled off site, transported by barge, and lifted into place. Sidewalks were placed outside the arch ribs on brackets supported by seven-and-a-half-foot floor beams.

3.12 Concrete and steel viaduct in foreground, steel arch in background. (Photo, F. S. Lincoln; courtesy of the New York Public Library)

THE VIADUCT

The bridge's exceptionally long approach rises at a steady 4 percent grade, so that the suspended roadway above the channel's main shipping lane offers clearance of 150 feet. A rigid economy in the design of the approaches was necessary to compensate for extra costs associated with sizing the main structure for the future addition of vehicular lanes and rail lines. To this end, the viaduct was designed with a road deck of ordinary steel-plate girder construction that spans reinforced concrete piers varying from 20 to 110 feet in height. The shorter piers consist of two independent hollow shafts; the taller piers are two hollow shafts connected at the top by an arch. Taller piers have footings that rest on solid rock, and the lower piers were built on spread footings. The concrete is reinforced with bolted steel frames to which formwork for the concrete was attached, with the structural steel remaining embedded in the piers. Steel rods near the surface provide secondary reinforcement.

3.13 Detail of the completed viaduct piers. (Photo, F. S. Lincoln; courtesy of the New York Public Library)

3.14 Construction of the steel-reinforced concrete piers of the viaduct. (Photographer unknown; courtesy of the Port Authority of New York and New Jersey)

Rather than forming the piers of solid concrete, the engineer made them hollow—not to strengthen them but to make them wider and therefore appear sturdier. Ammann believed a bridge should not only be strong, it should look strong. A keen concern for architectural aesthetics is evident in both the general proportions of the piers and in the crisp, understated detailing of their profiles.

THE LANGUAGE OF MONUMENTAL MASONRY

Ammann sought to translate the Hell Gate Arch Bridge's expression of monumentality to the longer and thinner Bayonne Bridge. The key in replication was the creation of enormous masonry abutments towering well above the ground. A pair of such massive abutments provided a perfect counterpoint to the light, lacy arch—an aesthetic juxtaposition, to Ammann. Furthermore, prominent abutments would bracket the arch, effecting a more strongly articulated transition between viaduct and suspended roadway.

From an engineering perspective, the abutments required to stabilize an arch on this site need barely rise above the ground. For architectural reasons, Ammann initiated a series

3.15 Study for the masonry surface designed for the ornamental abutments. (Rendering by George S. Dudley, Office of Cass Gilbert, Architect; courtesy of the New-York Historical Society)

opposite
3.16 Ornamental stonework was never erected above the bridge's abutments, leaving the steel armature exposed. (Photographer unknown; courtesy of the Port Authority of New York and New Jersey)

3.17 The arched truss viewed from the roadway. (Photographer unknown; courtesy of the Port Authority of New York and New Jersey)

3.18 The pedestrian walkway, cantilevered from the primary road deck. (Photo, F. S. Lincoln; courtesy of the New York Public Library)

of studies for an artificial superstructure that would rest on the structurally engaged section of the abutments. Cass Gilbert's office did a number of preliminary studies that played with variations in rustication and surface relief (fig. 3.15). All the schemes kept the height of the abutments even with the guardrail of the road deck. Although final plans for stone detailing were never released for fabrication, the steel framework that was built was designed to support the ornamental stone.

When the stone facade intended for the towers of the George Washington Bridge was eliminated by its supervisors, they made the decision well in advance of the bridge's opening. Supervisors of the Bayonne Bridge, however, simply and persistently put the order for the abutment's ornamental stone on hold, citing a need to exercise fiscal restraint. The governor of New Jersey, Morgan F. Larson, crowed at the opening ceremony, "it cost 14 percent less to build than was budgeted and appropriated."[2] Apparently the money *was* available to complete the approved design. When Ammann spoke at the opening ceremony, he lobbied for the ornamental abutments. Underscoring their formal significance, he told the assembled dignitaries, "The huge abutments of the arch, *which are yet exposed in their crude construction,* are eventually to be marked by massive pylons, and will thus further enhance the appearance of the structure in its setting in the landscape."[3] He and Gilbert continued to press for the stone after the bridge opened, and the engineering community sounded a number of well-placed appeals over the ensuing years. But the Port Authority never took action to either complete the construction or dismantle the spindly armature.

The Triborough Bridge

The Triborough Bridge is not at all a bridge in the ordinary sense of the term: it is an immense artery of travel in the very heart of New York City. The several distinct structures that make up the bridge connect three of the city's five boroughs while reaching out with its connecting highways to suburban developments at the metropolis's border. The enormous Y-shaped footprint of the facility is clearly visible from the air, but there is no single location on the Triborough roadway where all its components can be viewed together.

Of the structure's three individual long-span bridge structures, Ammann designed two: the Triborough Lift Bridge across the Harlem River and the Triborough Suspension Bridge over the East River. The third, the Triborough Truss Bridge crossing the Bronx Kill, was designed by Ash-Howard-Needles and Tammen, Consulting Engineers. Ammann also designed two and a half miles of connecting viaduct, including an elevated interchange and toll plazas above Randall's Island—a remarkably efficient and safety-conscious configuration that sorts out traffic flowing in twelve directions while providing access to the recreational facilities on the island itself.

The idea for the Triborough Bridge was first promoted by Edward A. Byrne, chief engineer of the New York City Department of Plants and Structures, in 1916. The merits of the proposal were widely hailed, but no progress was made until 1925 when the city appropriated funds for surveys, borings, and schematic structural plans. The appropriation was insufficient to complete the plan, so in 1929 an additional $3 million was allotted to cover the remaining planning work, construction of the Wards Island viaduct foundations, and the foundations for the towers of a suspension bridge designed by Arthur I. Perry. The city appropriated an additional $5 million in 1930, and contracts were awarded to construct the suspension bridge's anchorages. In 1932, with the suspension bridge's foundations and anchorages complete, city finances for the project dried up.

For two years, the anchorage and tower foundations stood in isolation. Convinced that the city would be unable to complete the work, the New York state legislature transferred

The growing importance of vehicular traffic and in particular the demand for accommodating fast through traffic with speed, comfort and safety has revolutionized the planning of major crossings. Notwithstanding their great size, these crossings have become mere links in a vast system of modern highway arteries and the planning of the approaches and highway connections has become one of the major problems of every new project.
—Othmar H. Ammann, 1945

the project from the Department of Plants and Structures to the Triborough Bridge Authority, under the chairmanship of Robert Moses, in the spring of 1933. Before construction resumed, Moses hired Ammann as chief engineer for the project. Ammann was already chief engineer for the Port Authority, and for six years he held that position in both authorities.

The various components of the comprehensive Triborough project—the purchase of land and easements, bridge and viaduct redesign and construction, expansion of existing streets and avenues into fourteen miles of approach highways and parkways, creation of parks and recreational structures, construction of new administrative office buildings for the Triborough Bridge Authority on Randall's Island—were financed through a combination of Depression-era federal relief funds (of which Moses was able to attract roughly $42 million through the Public Works Administration), municipal contributions, and revenues from the Triborough Bridge Authority, including bond sales. Bridge tolls would meet bond and other debt obligations and finance maintenance of the facility. The Triborough Bridge Authority (later renamed the Triborough Bridge and Tunnel Authority, and now

known as MTA Bridges and Tunnels) was to handle all aspects of the project: financing, construction, and facility operation.

The project created the Depression-relieving jobs its political backers in New York City and Albany had promised. Building the Triborough employed an average of a thousand men in the field daily, reaching peak employment in June 1936, when twenty-eight hundred workmen were on the site. Many times more worked at manufacturing materials, factory assembly, shipping, and the like. In all, workforces in 134 cities scattered across twenty states participated.

The Triborough Bridge was to provide commercial traffic with a seamless path through and around the north-central quadrant of the metropolitan area. Private motorists figured prominently in the planning, too. Because of the Triborough, Long Island residents could reach upper Manhattan, the Bronx, Connecticut, New England, and the west without crossing either the heavily traveled Queensboro Bridge or the Long Island Sound by way of ferry. Those going from any of those places to Long Island enjoyed the same benefit. As for Manhattan residents who wished to travel north or east, the bridge and its connections offered a whole new escape route. After struggling through the traffic-jammed streets of Manhattan to Ninety-second Street and the East River, drivers were able to move with virtually no interruptions along East River Drive—built as part of the Triborough project—to the Triborough Bridge, where convenient connections were available to highways leading to Westchester and Connecticut. These connections were further simplified later with the construction of the Bronx–Whitestone and Throgs Neck Bridges and their connecting highways.

THE TRIBOROUGH LIFT BRIDGE

Many critics consider the Triborough Lift Bridge the most beautiful structure in the Triborough complex. Certainly it is among the most architecturally refined lift bridges ever to have been built. It crosses the Harlem River at East 125th Street by means of three independent truss spans: one short span at either end, reaching to the towers, and a 310-foot movable central span. In the down position, the central span has a 55-foot underclearance, which serves most river traffic, but it can be raised to a position 135 feet above water level, offering ample clearance for high-masted vessels.

The movable truss carries a single-deck roadway with four traffic lanes. It hangs from the towers by ninety-six wire ropes wound around fifteen-foot-diameter drums; hoisting

4.3 With the Hell Gate Bridge at its side, the four sections of the Triborough Bridge are clearly seen in this photograph, circa 1948. The Lift Bridge is at the upper left; the Suspension Bridge at lower right; the Truss Bridge designed by Ash-Howard-Needles and Tammen, Consulting Engineers, at top center; and the Triborough Viaduct over Wards and Randall's Islands ties them all together. (Photographer unknown; courtesy of MTA Bridges and Tunnels, Special Archives)

4.4 Detail of the Triborough Lift Bridge. (Photo, Berenice Abbott, circa 1937; courtesy of the New York Public Library)

opposite

4.5 The Triborough Lift Bridge in the foreground, the Triborough Viaduct and Hell Gate Arch Bridge in the background. (Photo, F. S. Lincoln; courtesy of the New York Public Library)

4.6 Pedestrian approach to the movable roadway.

(Photo, F. S. Lincoln; courtesy of the New York Public Library)

4.7 Vehicular approach from East 125th Street to the movable roadway.

(Photo, Berenice Abbott, 1937; courtesy of the New York Public Library)

top left

4.8 The Manhattan approach to the Triborough Lift Bridge gathers traffic from First Avenue and East River Drive, 125th Street, and Willis Avenue in the Bronx by way of the Willis Avenue Bridge (foreground). Ammann's Harlem River Pedestrian Bridge is visible in the background, center. (Photo, Jet Lowe; courtesy of the Historic American Engineering Record, National Park Service)

top right

4.9 The trussed road deck, looking east. (Photo, Berenice Abbott; courtesy of the New York Public Library)

bottom left

4.10 The Triborough Lift Bridge, with its circular approach ramp from East River Drive; beyond it is the first section of the Triborough Viaduct with its east/westbound toll and exchange plaza elevated above Randall's Island. Off ramps near the toll booths provide access to parks on Randall's and Wards Islands. (Photo, Jet Lowe; courtesy of the Historic American Engineering Record, National Park Service)

bottom right

4.11 Aerial view of the north/southbound exchange and toll plaza. A section of the Triborough Truss Bridge appears in the foreground. (Photo, Jet Lowe; courtesy of the Historic American Engineering Record, National Park Service)

the 2,050-ton roadway involves turning the drums to reel in the suspension ropes. Power to operate the drums comes from four two-hundred-horsepower electric motors, housed in crenellated steel cases at the top of the towers, and a thousand-ton concrete counterweight that moves up and down inside a tower leg.

The general outline of the 210-foot towers recalls those of the George Washington Bridge. Two sets of braced four-column legs compose each tower. The two groups of braced legs are in turn braced at their tops by a 20-foot-deep truss that forms an arched opening. Deviating from the standard engineering practice of resting the back leg of the supporting tower on the adjacent side span, here each leg rests on its own concrete pier, a move Ammann made to improve the appearance of the bridge.

When the Triborough Lift Bridge opened, its deck area, at twenty thousand square feet, was the largest of its type. It was not, however, the heaviest span, by dint of clever design. Ammann made every effort to reduce its weight and, by extension, cost; rather than pave the roadway with concrete, for example, much lighter paving planks of asphalt were laid on the steel-plate road-deck girders.

As with the Bayonne Bridge, the Triborough Lift Bridge's sidewalks are cantilevered from the deck girders. They are part of a continuous pathway that runs through the complex, connecting pedestrians and bicycles to park facilities on Randall's and Wards Islands and to neighboring boroughs.

THE TRIBOROUGH VIADUCT AND RANDALL'S ISLAND INTERCHANGE PLAZA

Each arm in the Triborough Viaduct Bridge is linked to one of the three bridge spans: the Triborough Lift Bridge, the Triborough Suspension Bridge, or the Triborough Truss Bridge. The viaduct's three arms converge on an enormous traffic-sorting, toll-gathering interchange plaza elevated over Randall's Island. More than two and a half miles long all told, the viaduct travels 1,570 feet over Queens, 3,000 feet over Wards Island, 4,750 feet over Randall's Island, 3,500 feet over Manhattan, and 1,000 feet over the Bronx.

Most of the viaduct's eight-lane roadway is supported by steel-plate girders resting on concrete piers 60 to 140 feet apart. The roadway's surface is built with a network of I-beam stringers that support steel cross beams overlaid with concrete slabs. Operators in cranes working on the ground lifted the steel for the Queens and Manhattan approach viaducts. On Wards Island, Randall's Island, and over the Little Hell Gate, where the roadways are

4.12 Looking northward along the Triborough Viaduct to the north/south exchange plaza and the Triborough Truss Bridge. The water channel that separated Randall's from Wards Island has been filled with earth. The former course of the channel is marked by the presence of the Little Hell Gate Bridge (the small bridge between the Triborough Viaduct and the Hell Gate Bridge at right) and the change in the Hell Gate Bridge's viaduct structure from closely spaced concrete piers designed to rest on land to four trussed spans on widely spaced concrete columns, which carried the rail deck across the water. (Photo, Jet Lowe; courtesy of the Historic American Engineering Record, National Park Service)

below left

4.13 Girder installation for the Triborough Viaduct. (Photographer unknown; courtesy of MTA Bridges and Tunnels, Special Archives)

below right

4.14 Viaduct construction on Wards Island, with the Triborough Suspension Bridge rising in the distance. (Photographer unknown; courtesy of MTA Bridges and Tunnels, Special Archives)

at a much greater height above the ground, a movable crane mounted on the deck girders did the work.

The piers that support the roadway are made of three columns joined at their tops with reinforced concrete ties in the form of arched soffits. The columns are octagonal in cross section—a design proposed on aesthetic grounds that proved to also offer a savings in material over a rectilinear shaft. The detail at the piers' summit is a simple ornamental banding developed by Ammann's consulting architect, Aymar Embury II, who had also proposed the octagonal column shape.

The viaduct's roadway widens as it approaches the interchange plaza to accommodate toll collection, which occurs at two points: vehicles passing through Queens and the Bronx pay tolls where the two roadways of the main viaduct are combined and widened to 137 feet; all others pay tolls in an area on the Manhattan branch where the viaduct is widened to 195 feet. The plaza proper has a roadway area of 390,000 square feet (approximately nine acres), supported on more than 1,700 concrete columns, and is faced on all sides by a concrete curtain wall 8,000 feet long. In all, the plaza contains 70,000 cubic yards of concrete and 5,900 tons of reinforcing steel (almost as much steel as is contained in the two towers of the Triborough Suspension Bridge).

The plaza is surrounded by two concentric circular ramps arranged to preclude grade crossings and left turns and to orchestrate toll entrance but free exits. Ammann stated that the interchange plaza combined "speed and safety as well as a form of recreation to the traveling public to an extent not equalled in any similar project."[1] Few people today are likely to consider passage through the interchange plaza a form of recreation, and despite the best efforts of the traffic engineers, navigating the interchange really can be confusing. Many motorists have innocently taken a wrong turn and found themselves speeding away from their destination rather than toward it. Their only recourse is to retrace their steps, probably pay another toll, and hope that signage and their sense of direction will serve them better on their second loop through the plaza.

THE TRIBOROUGH SUSPENSION BRIDGE

Casual observers commonly refer to the Triborough Suspension Bridge as the Triborough Bridge per se. The mistake is understandable, as the suspension bridge is the most conspicuous feature in the complex, and because the individual structures have never been given formal names. The Triborough Suspension Bridge is the tallest and longest span in

4.16 Raising the first segment of the road deck's stiffening truss. (Photographer unknown; courtesy of MTA Bridges and Tunnels, Special Archives)

opposite
4.15 The Triborough Suspension Bridge as seen from above Astoria in Queens. (Photo, Jet Lowe; courtesy of the Historic American Engineering Record, National Park Service)

the complex. Its 1,380-foot clear span is comparable to those of other East River bridges: the Queensboro Bridge, with a clear span of 1,182 feet; the Manhattan Bridge, which spans 1,470 feet; and the Brooklyn and Williamsburg Bridges, with spans of 1,595 and 1,600 feet, respectively.

The Triborough Suspension Bridge, as originally designed by Arthur I. Perry, had its groundbreaking on the morning of October 25, 1929, inauspiciously, as it turned out. Later that day, prices on the New York Stock Exchange crashed, signaling the onset of the Great Depression. The anchorage and tower foundations were completed in the months that followed, but the city's bleak economic climate, combined with a series of cost overruns, forced officials to prematurely stop construction.

When the project was revived in 1933, Ammann was engaged to revise Perry's design, because estimates for its completion greatly exceeded appropriations. Ammann's redesign resulted in considerable savings (approximately $10 million in the towers alone), largely from the simple scaling back of the facility's size. The original roadway was to be double-decked, with eight lanes per deck. The redesign eliminated one deck, leaving a single eight-lane roadway stiffened with a Warren truss. This change permitted a reduction in the number and size of other bridge components: Ammann's towers stand on two legs, for example, where the original had four, and his structure has two suspension cables instead of four.

During the 1930s, Ammann demonstrated a preference for plate-girder road decks over those stiffened by Warren trusses. But because he had to develop the Triborough Suspension Bridge design from existing tower foundations and anchorages configured for a deep, double-decked roadway, he worked Warren trusses into the redesigned road deck. The deck's floor system does contain some remarkable plate girders, though, functioning as floor beams, ninety-six feet long (the width between the Warren stiffening trusses) and eight feet four inches deep (fig. 4.17).

Cranes lifted these deck members into place for construction and assembly. Only half the structure was raised during the first sweep of the traveling crane. (Had the full deck structure been assembled, its weight would have thrown the sag in the suspension cables off balance.) Once the first layer was in place, the operator redirected the crane back toward the originating towers, filling in the missing pieces as he returned.

The roadway is tucked between the top and bottom chords of the stiffening trusses, partially obscuring views of the surrounding landscape for motorists. But walkways cantilevered over the water from the top chord of the stiffening trusses offer pedestrians open vistas to the river and the moving traffic on the bridge.

4.19 Contrary to the design of his other suspension structures, Ammann ran the Triborough's suspension cables over the saddles of a cable-bent post, changing the angle of inclination of the cables before they reach the eyebars in the reinforced-concrete anchorage. (Photographer unknown; courtesy of MTA Bridges and Tunnels, Special Archives)

top
4.20 Construction detail of the cable-bent post. (Photographer unknown; courtesy of MTA Bridges and Tunnels, Special Archives)

bottom
4.21 Profile of the anchorage system during construction. (Photographer unknown; courtesy of MTA Bridges and Tunnels, Special Archives)

The Triborough Suspension Bridge, in another vestige of the first construction phase, has exposed cable bents—the spot where the compacted suspension cable changes its course from practically horizontal to forty-five degrees. Part of the force in the cable is carried to the ground at this spot through the strut post of the bent. The remaining force is transferred to the heel of the anchorage, where the wires of the splayed cable are attached to eyebars embedded in concrete.

Embury, the consulting architect, developed the architectural treatment of the anchorages with a shape suggesting the dynamic balance between the direction of the cable's pull and the necessary redistribution of weight to counteract overturning. The anchorage surfaces are scored with V-shaped cuts that correspond to the angles of the cables as they are splayed within the anchorages. These cuts also cleverly serve to conceal expansion joints that minimize and control cracking along the surface of the monolithic concrete.

TOWERS

Much as when he had been thwarted in his intention to enclose the open steelwork in the towers of the George Washington Bridge with stone, Ammann found himself in another kind of compromised position when he assumed the role of chief engineer for the Triborough. Perry had already designed towers for the suspension bridge that roughly translated the stone profiles of the Brooklyn Bridge towers into steelwork tracery (fig. 4.24). Several of Perry's tower components had been manufactured before the project was temporarily shut down. When it started up again, in the interest of fiscal austerity, nothing was to be wasted. Ammann therefore faced the daunting challenge of reconfiguring the towers while using remnants from the former design.

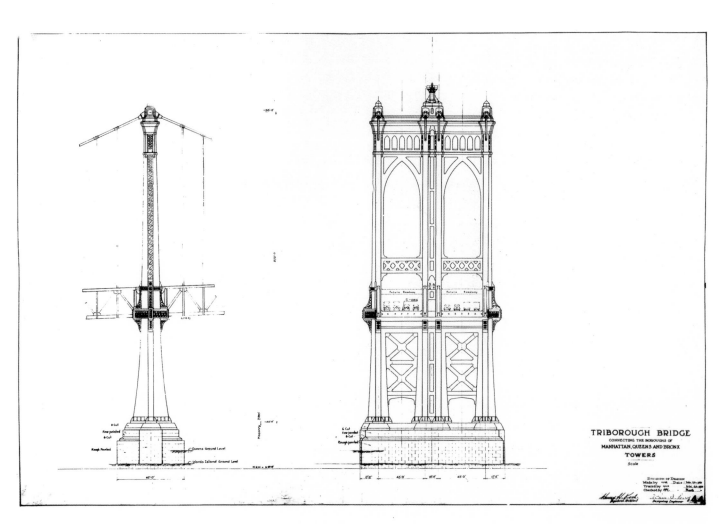

4.24 Side and front elevation of Arthur I. Perry's original design for the Triborough Suspension towers. (Courtesy of MTA Bridges and Tunnels, Special Archives)

Ammann's tower design reflects his aesthetic preferences as well as programmatic changes in the bridge. His towers are lighter, crisper, and considerably less romantic than Perry's, and they follow the modernist aesthetic of simple geometry and uncluttered lines (fig. 4.23).

The tower legs are of cellular construction, consisting of two cruciform-shaped verticals connected by cross-bracing below and above the roadway and across the top. (The bases of the legs were fabricated for the Perry design). The legs use silicon steel and the bracing is of carbon steel; together the towers incorporate a total of 5,500 tons of steel. The four ninety-three-foot posts that link the top of the intermediate cross-brace with the bottom of the uppermost cross-brace are ornamental: serving no structural purpose, their function is simply to emphasize the vertical. The strut at midheight steadies the posts in the wind.

All sections of the tower were prefabricated in an off-site factory. Once on site, a crane lifted them into position and workers bolted them in place. The tower-assembly operation used only one crane, which lifted the steel for each tower in three phases, with the two tow-

left

4.25 The central section of the lowest strut being lowered into place between the legs of the tower. (Photographer unknown; courtesy of MTA Bridges and Tunnels, Special Archives)

right

4.26 Raising a segment of the middle truss that serves as a strut for the tower. (Photographer unknown; courtesy of MTA Bridges and Tunnels, Special Archives)

4.27 Shop fabrication of the middle tower strut. (Photographer unknown; courtesy of MTA Bridges and Tunnels, Special Archives)

4.28 Looking down the vertical posts that connect the top and middle struts of the tower. (Photo, Dave Frieder; courtesy of Dave Frieder)

ers going up in staggered fashion. With each completed phase, the crane was disassembled and moved to the other tower to build it to the next level. This sequence proved ideal from a safety standpoint because it separated the operation of erecting and bolting from riveting: workers riveted one tower while the other was being built. Remarkable accuracy in the shopwork on the preassembled tower components brought a new level of efficiency to the construction; the milling of the column sections was so close to the specified lengths that it was unnecessary to follow the customary practice of making special corrections in the field to the top sections.

At the summit of the towers, the suspension cables run over cast-iron saddles where the wires were spun and laid. The saddles absorb movement in the cable caused by loading and temperature changes and deflect it to the towers. Thirty-foot-tall architectural lanterns surmounted with aviation lights provide a whimsical touch at the top of each saddle (fig. 4.22).

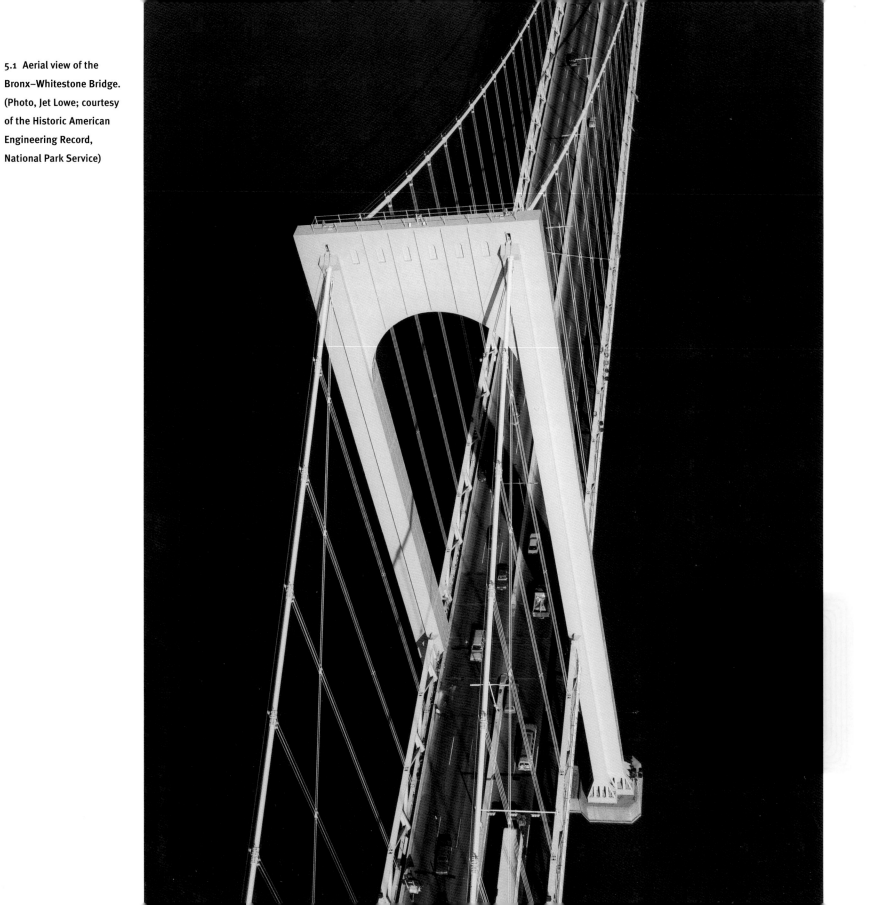

5.1 Aerial view of the Bronx–Whitestone Bridge. (Photo, Jet Lowe; courtesy of the Historic American Engineering Record, National Park Service)

The Bronx–Whitestone Bridge

As originally constructed, the Bronx–Whitestone Bridge is a technical tour de force and an aesthetic masterpiece of unprecedented simplicity and slenderness. At 2,300 feet, the bridge had the fourth-longest clear span in the world when it opened in 1939, and the public widely appreciated its beauty and utility. But public perceptions of its safety began to sour a year after it opened, for the road deck occasionally moved perceptibly. Ammann judged the movement harmless, but the public did not, and the resulting concern precipitated a retrofit installation of Warren trusses as stiffeners, undermining the splendor of the original structure's appearance.

Ammann developed the ribbonlike road deck with the assistance of design consultant Leon S. Moisseiff. Experimental in nature, the deck was engineered with nothing more than a pair of eleven-foot-deep steel plate girders that maintain their profiles from anchorage to anchorage; all other parts of the structure supporting the roadway fit within the girders' depth. A line in space could not be more economically drawn. In addition to the bridge's beauty from a distance, travelers passing over it enjoyed unobstructed views of splendid scenery because the tops of the stiffening girders stayed below generous sight lines.

With its novel towers of plate steel, no suspension bridge had ever been better proportioned or more refined, reduced, and elongated than the Bronx–Whitestone. At the official ribbon cutting, Robert Moses described it as "architecturally the finest suspension bridge of them all, without comparison in cleanliness and simplicity of design, in lightness and absence of pretentious ornamentation. Here, if anywhere, we have pure, functional architecture."[1]

The bridge was completed in time to carry motorists to New York's Depression-era extravaganza, the 1939 World's Fair. Although construction of the bridge was tied to the vigorous building campaign associated with the fair, regional planners had foreseen a need for a bridge on the site nearly ten years before promotion of the World's Fair began. A Bronx

It is only with a broad sense for beauty and harmony, coupled with wide experience in the scientific and technical field, that a monumental bridge can be created.
—*Othmar H. Ammann, 1918*

crossing to the Whitestone section of Queens was part of a circumferential highway program developed between 1921 and 1928 by the Regional Plan Association, which considered it necessary to carry the Belt Parkway from the northern end of the Cross Island Parkway through Queens and across the East River into the Bronx, where it would connect with Eastern Boulevard and the southern extension of the Hutchinson River Parkway, leading to Westchester County, Connecticut, and upstate New York. This link would also provide a means of reaching southern and eastern Long Island from both the south and the north without passing through the central business district or high-density residential areas in Queens.

Although the bridge ignited a local housing boom in parts of Queens, its greatest impact was regional in scale. Like the Triborough, Bayonne, and George Washington Bridges, the Bronx–Whitestone had a "tiers only" policy (decks were not engineered to accommodate rail). Thus, as the metropolitan population began to spread out from its long-established core in Manhattan, the bridge reinforced trends away from nuclear development centered at stops on a rail network and toward private automobile ownership and the establishment of low-density residential districts broadly scattered across the region. The

5.2 View from Francis Lewis Park, in the Whitestone section of Queens. The designer's drive for clarity and economy are reflected even in the anchorages: the eleven-story structures are exposed concrete shaped and grooved to carry the curving line of the main suspension cable to the ground without undue formal interruption. (Photo, Dave Frieder; courtesy of Dave Frieder)

5.3 As originally constructed, the Bronx–Whitestone epitomized Ammann's ideal of formal elegance: the lines are reduced in number, simplified, and highly elongated. (Photo, Richard Averill Smith; courtesy of MTA Bridges and Tunnels, Special Archives)

highways were changing the industrial landscape too, allowing the trucking industry to fully participate in a burgeoning regional trade supported by New York's harbor, railways, and, increasingly, airports. As with housing, industry sprawled along the seemingly boundless highways.

5.6 Approach ramp leading to the main span, illustrating the original lighting and landscape features. The rustic wooden lampposts and guardrails reinforced the pastoral character of the parkway to the bridge. (Photo, Richard Averill Smith; courtesy of MTA Bridge and Tunnels, Special Archives)

5.7 View from midspan of the original roadway configuration. (Photo, Richard Averill Smith; courtesy of MTA Bridges and Tunnels, Special Archives)

5.8 Francis Lewis Park—built as an extension of the parkway system serviced by the Bronx–Whitestone Bridge—offers a spectacular view of Ammann's structure. This photograph, taken in July 1941, shows the diagonal cable bracing that was added after the bridge opened. (Photo, Rodney McKay Morgan; courtesy of MTA Bridges and Tunnels, Special Archives)

STRUCTURAL MODIFICATIONS

At the time the Bronx–Whitestone Bridge was on the drawing board, the engineering community was largely unaware of the destabilizing effect wind could have on extremely lightweight and slender long-span suspension bridges. Although Ammann predicted that the Bronx–Whitestone's shallow road deck would move more dynamically under the force of wind than one stiffened with conventional Warren trusses, a saving of $2 million accompanied the choice of the elegant plate-girder stiffening system. An occasionally perceptible level of movement seemed a small price to pay for both beauty and a financial bargain.

Moisseiff, Ammann's consultant for the project, had designed a long-span bridge in Tacoma, Washington, that opened to traffic early in 1940, about six months after the New York bridge. Moisseiff's Tacoma Narrows Bridge closely followed the design of the Bronx–Whitestone, though it was longer and narrower. When the Tacoma bridge's roadway was torn to shreds under the force of a relatively mild wind on November 7, 1940, the spectacular disaster ignited public panic.

The Tacoma Narrows Bridge had been given the appropriate nickname "Galloping Gertie" even before it failed, because of the road deck's demonstrated instability during construction. The "gallops" or rolling oscillations that preceded the bridge's collapse created tremendous torsional stresses in the deck. It lacked sufficient stiffness to stand firm against the power of the wind, and the farther it moved up and down, the greater the stress. Once set in motion, the steady wind accelerated the gallops.

In response to public concern over the safety of long-span bridges, the federal government convened a blue-ribbon commission, which Ammann headed, to investigate the Tacoma Narrows failure. The commission determined that a number of factors had contributed to the bridge deck's excessive flexibility. It was too lightweight: the central span of 2,800 feet was 500 feet longer than the Bronx–Whitestone, but its suspended deadweight per foot was less than half that of the New York bridge. The deck was also too shallow: its solid-web stiffening girders were only eight feet deep, or 1/350 of the center span; the Bronx–Whitestone's eleven-foot deck was 1/209 of the center span. Furthermore, the Tacoma Narrows side spans were long relative to the clear span, and the cables were anchored a considerable distance beyond the side spans: a more favorable interval would have tempered flexibility. Finally, the roadway's width relative to the clear-span length represented an unusually low ratio: 1:72 for the Tacoma Narrows Bridge, whereas the ratio for the George Washington Bridge is 1:33 and the Bronx–Whitestone Bridge 1:31.

5.9 The Bronx–Whitestone Bridge shown with its retrofitted trusses on the road deck, as well as the diagonal bracing added earlier to the suspension system. (Photographer unknown; courtesy of MTA Bridges and Tunnels, Special Archives)

5.10 Detail of the Bronx–Whitestone's retrofitted profile.
(Photo, Jet Lowe; courtesy of the Historic American
Engineering Record, National Park Service)

Moisseiff had engineered his bridge in complete compliance with accepted theory and practice. The failure dramatically revealed a serious flaw with the deflection theory as it had conventionally been applied. Previously, engineers had focused on stiffening long-span bridges against the destabilizing effects of moving traffic. It was assumed that if they addressed this form of dynamic load, stiffness would also be adequate against what they considered the more minor effect of wind. The Tacoma Narrows failure demonstrated that long, light, suspended road decks will behave somewhat like the wing of an airplane. The Tacoma commission's seminal conclusion was that aerodynamics must become a primary engineering consideration in future lightweight, long-span design.

Ammann was aware that the Bronx–Whitestone Bridge's deck noticeably oscillated from time to time—it had first done so while it was under construction. Although it was perfectly harmless to the structure, the rare occasions when oscillation was amplified undermined the comfort level of people walking or driving across the bridge. In an effort to enhance the deck's stability, a number of minor alterations were made to the structure in 1940, most notably the addition of diagonal stiffening cables running from the tower tops to the plate girders. Understandably, accounts of the Tacoma Narrows failure raised new concerns about the aerodynamic stability of the Bronx–Whitestone. Motorists who regularly crossed the bridge were particularly vocal in their demand for a structural reassessment. After statistical reexamination—and subsequent wind tunnel testing—Ammann determined that alterations to the bridge were unnecessary, that it would be stable in any foreseeable wind conditions.

Public perceptions outweighed hard science. The chairman of the Triborough Bridge and Tunnel Authority, Robert Moses, gave orders to further stiffen the bridge, and in 1946 Ammann oversaw the retrofitting of two Warren trusses fourteen feet tall that were connected to the tops of the Bronx–Whitestone's plate girders; the trusses dip at midspan to accommodate the main suspension cables. At the time of the retrofit, the bridge's sidewalks were eliminated and its six traffic lanes widened. Lost forever were the pure and simple lines of the original structure and the unobstructed views its roadway afforded travelers.

About the Tacoma Narrows collapse, Ammann later reflected, "regrettable as the Tacoma Narrows Bridge failure and other recent experiences are, they have given us invaluable information and have brought us closer to the safe and economical design of suspension bridges against wind action."[2]

The Throgs Neck Bridge

Between 1931 and 1939, four long-span bridges opened to motorists in the New York metropolitan area—all designed by Ammann. More than twenty years elapsed before the completion of the next major bridge project in the region, the Throgs Neck Bridge in 1961. Population and flow of trade had exploded during the 1940s and 1950s, straining the capacity of the bridges and tunnels in the metropolis. In an effort to address mounting pressures on the highway system, the Port Authority of New York and New Jersey, together with the Triborough Bridge and Tunnel Authority, released a planning study in 1954 that made among its recommendations the installation of a second deck to the George Washington Bridge, the building of a bridge across the Verrazano Narrows, and the construction of the Throgs Neck Bridge. Its six vehicular lanes were needed to relieve congestion on the Bronx–Whitestone Bridge two miles west.

Unlike Ammann's other major projects, the Throgs Neck Bridge was not an element in the circumferential highway system laid out in 1928 by the Regional Plan Association. No one in 1928 could have envisioned the phenomenal ascendancy of the motor vehicle over a thirty-year period that witnessed industrial innovations attending the Second World War, followed by President Dwight D. Eisenhower's expansion of the interstate highway system. The annual volume of vehicles crossing the Triborough Bridge had steadily increased from 13 million in 1946, the first year after World War II, to 32 million in 1951. The Bronx–Whitestone Bridge, which had been built in part to relieve traffic congestion on the Triborough, had experienced a similar growth in traffic. Overall, traffic had increased 138 percent on the Triborough Bridge between 1946 and 1951 and 129 percent on the Bronx–Whitestone Bridge. By 1960, the Bronx–Whitestone was carrying 33.2 million vehicles annually. When the Throgs Neck Bridge first opened, traffic on the Bronx–Whitestone dropped by 40 percent. By 1966, however, both facilities were accommodating more than 30 million vehicles a year, illustrating a now widely held view among transportation plan-

It is true that aesthetic conceptions had to adjust themselves to the new materials and structural forms. Architectural embellishments, so admirably exemplified in many smaller, older bridges, have no place in modern large bridges. On the other hand, there is no justification for artificial or hybrid designs which crop up from time to time on the ground of scientific expedient. Pleasing appearance must be produced by a clear expression of the natural function of the structure; and by simple, pleasing lines and proportions.

—Othmar H. Ammann, 1954

ners: any increase in a heavily traveled highway network's capacity is quickly followed by a surge in traffic levels that reestablishes the network's crowded conditions.

Built and operated by the Triborough Bridge and Tunnel Authority, the Throgs Neck Bridge's suspended roadway hovers over the confluence of the East River and Long Island Sound before passing over the grounds of the New York State Maritime College at Fort Schuyler in the Bronx, where it comes to rest in a sweeping curve across a shallow bay. More than two miles of over-water viaduct connect with the suspended roadway. On the Queens side of the bridge, land acquisition for the highway involved the displacement of 421 homes in the comfortably middle-class neighborhood of Bayside. Most of these homes were lifted from their foundations and relocated to a new neighborhood built on the site of the Bayside and Oakland golf courses. To build on the Bronx side, the Triborough Bridge and Tunnel Authority had to obtain a right-of-way from the New York State Maritime College at historic Fort Schuyler. The college granted the easement in return for an adjacent landfill project on their property.

Interestingly, three of Ammann's bridges overlap the site of historic naval fortifications: Fort Washington (established in 1776) in Manhattan at the George Washington Bridge; Fort Wadsworth (1663) on Staten Island and Fort Hamilton (1831) in Brooklyn at the Verrazano-Narrows Bridge; and Fort Schuyler (1856). The two underlying site conditions that define an ideal location for a naval fortification are identical to those for a suspension bridge: ground conditions must be solid and compact to properly support foundations for the massive masonry walls of a fort or the enormous towers of a suspension bridge, and the narrower the channel at the foot of the building site the better. For an artillery-mounted fortification, a narrow channel means a shorter firing range. Narrow channels are ideal for a suspension bridge because they permit short, economic spans.

A BRIDGE WITH MUSCLE

Uncompromised in its design, the Throgs Neck Bridge is a perfect exemplar of suspension's inherent elegance. Unlike the other New York bridges Ammann engineered, the Throgs Neck commission posed no constraints that forced the designer to push the boundaries of established practice (its eighteen-hundred-foot clear span is five hundred feet shorter than that of the Bronx–Whitestone). Ammann, therefore, proposed a structure based on well-established conventions. This is not to suggest that he had grown timid. While finishing the Throgs Neck, he was developing the design for what was to be the next reigning "longest suspension bridge in the world"—the Verrazano-Narrows Bridge.

The form of the Throgs Neck Bridge approximates that of the retrofitted Bronx–Whitestone, although the twenty-eight-foot-deep stiffening trusses in the deck of the newer bridge are kept below the eye level of motorists (fig. 6.3). Compared with the Bronx–Whitestone, the Throgs Neck is blockier in its proportions and stiffer in form (though considerably shorter in span, the Throgs Neck towers are only twenty-two feet shorter than the Bronx–Whitestone's). Although both bridges' towers are built of two steel columns of closed box construction tied together at the top and just below the road deck with arched struts, the struts of the Bronx–Whitestone towers form portals possessing truly seamless lines made with hemispheric struts, whereas the struts of the Throgs Neck present the less fluid line of a flattened segmental arch. The Throgs Neck's posture is taut and muscular.

6.3 Profile of the bridge illustrating the relationship of the deck truss to the roadway surface. (Photo, Jet Lowe; courtesy of the Historic American Engineering Record, National Park Service)

6.4 The road deck was assembled with large truss segments hoisted into position beginning at the towers. (Photographer unknown; courtesy of MTA Bridges and Tunnels, Special Archives)

6.5
View of the underside of the road deck structure. (Photo, Dave Frieder; courtesy of Dave Frieder)

6.6 The Throgs Neck Bridge, foreground, with the Bronx–Whitestone Bridge behind it.

(Photo, Jet Lowe; courtesy of the Historic American Engineering Record, National Park Service)

The bridge's deck structure is exceptionally stable. It carries six lanes of traffic that rest on a series of laterally arranged transverse floor trusses (fig. 6.5). A barricade four feet wide divides the roadway into two thirty-seven-foot vehicular corridors, flanked by two-and-a-half-foot emergency sidewalks adjacent to the perimeter barricade and railings of steel. The transverse floor trusses are framed into two longitudinal stiffening trusses located in the vertical planes of the suspension cable (the stiffening trusses form the perimeter barricades). There is a lateral system of stiffening trusses in the horizontal planes of the top and bottom chords of the floor truss to brace the deck against the wind. Together, the floor trusses and the stiffening trusses form a rigid frame that offers ample resistance vertically and torsionally against dynamic wind action. The effects of wind action are further minimized by the space left between the pavement and the topmost lateral stiffening trusses. The pavement on the roadway consists of bituminous concrete resting on a five-inch-deep slab formed by steel-grid flooring filled with concrete. These slabs rest directly on continuous longitudinal stringers supported on the top chords of the transverse floor trusses.

The primary structural components of the deck were fabricated in sectional units off site, brought to the bridge by barge, and lifted into place by crane. Raising the deck began at each tower and proceeded simultaneously toward the midspan and the approach roads, the same sequence followed in the construction of the Triborough Suspension and Bronx–Whitestone Bridges. By contrast, assembling the prefabricated deck components for the Verrazano-Narrows Bridge began at midspan and proceeded toward the towers.

opposite

6.7 View to the road deck from the top of the Queens tower. (Photo, Dave Frieder; courtesy of Dave Frieder)

7.1 The Verrazano-Narrows Bridge, looking down from the main cables to the portal opening of the tower. (Photo, Dave Frieder; courtesy of Dave Frieder)

The Verrazano-Narrows Bridge

The Verrazano-Narrows Bridge is Ammann's crowning achievement—a suspended structure with a clear span nearly one mile in length. It is a landmark gateway to New York Harbor, the last arterial link forged in the metropolis's circumferential highway system, and to this day the only highway connection between Staten Island and another New York City borough.

New York has witnessed a long history of failed efforts to cross the Narrows. In 1888, the Baltimore & Ohio Railroad attempted to tunnel the Verrazano Narrows. The company had built a swing bridge from New Jersey to Staten Island several years earlier but was still using ferries to carry goods from Staten Island to Brooklyn and Manhattan. The planned tunnel would have greatly expanded their commercial interests, but the initiative was stalled by delays in the approval process and difficulties with financing. It advanced no further than the drawing board.

In 1923, New York mayor John F. Hylan championed a crossing for the sake of Staten Island's general development. He successfully secured the considerable sum of $500,000 in appropriations to begin a combination freight and passenger tunnel. Crews broke ground but the project was soon abandoned, leaving two uncompleted tunnel shafts—one in Bay Ridge, Brooklyn, and one near Fort Wadsworth on Staten Island. The excavations came to be sarcastically known as "Hylan's Holes."

In 1926, New York engineer David B. Steinman proposed a suspension bridge across the Verrazano Narrows. At the time, the longest suspension span in the world—the recently opened Philadelphia–Camden Bridge (now known as the Benjamin Franklin Bridge) designed by Ralph Modjeski—was a mere 1,750 feet long. Steinman's audacious "Liberty Bridge" between Brooklyn and Staten Island would have a 4,620-foot clear span. The steel towers of Steinman's proposed structure were to be 800 feet tall and crowned with Gothic tracery enclosing observation decks, beacon lights, and a clarion of bells. A persuasive entrepreneur, Steinman talked a group of private investors into forming a business syndicate

This span and clearance arrangement, combined with the structurally and economically required general dimensions of the structure, the width and depth of the double deck, the sag of the cables, and the height of the towers, resulted in a well-balanced design of light and graceful appearance, in harmony with the magnificent landscape. The bridge will form a monumental portal at the entrance of the New York Harbor.
—Othmar H. Ammann, 1963

7.2 The running of the New York Marathon is among the city's most celebrated annual events. The race starts on the Staten Island side of the Narrows, and crosses the bridge to get to Brooklyn (shown here). (Photo, Barton Silverman/NYT Pictures; courtesy of the *New York Times*.

and applying to Congress for a charter to build and operate the facility. The charter would have been granted had it not been for Congressman Fiorello H. La Guardia, who single-handedly blocked a vote on the measure. La Guardia was not opposed to a bridge per se: he just believed that private capital should not profit at the expense of civic necessity, and that a crossing should be owned and operated by a public authority. When he later served as mayor of New York, La Guardia campaigned for a Narrows crossing—in part because such an ambitious public work would help relieve regional unemployment, presumably at the federal government's expense.

With the Steinman proposal defeated, in 1929 the New York City Board of Transportation took up the cause of a vehicular tunnel beneath the Narrows, but later that year the

Great Depression took hold. The economic shock waves set off by the Depression stifled confidence in new ventures, and again no action was taken beyond the development of a planning study. Twelve years later, near the end of the Depression, the City Planning Commission adopted the "Master Plan for Arterial Highways and Major Streets," which included a Narrows crossing, but the La Guardia administration failed to mount an adequate fund-raising campaign to implement construction. While the administration was haplessly struggling in the mid-1940s to build a crossing, the chairman of the Triborough Bridge and Tunnel Authority (TBTA), Robert Moses, began his own drive to build a transportation link at the Verrazano Narrows.

First, Moses had plans drawn for a tunnel. To his apparent satisfaction, the planning study showed that a two-lane tunnel running under the Narrows' bedrock would cost more than a two-level, twelve-lane suspension bridge. Taking up the cause for a bridge, Moses began to shepherd the project through an arduous twelve-year approval process.

In 1949, the U.S. Department of the Army granted Moses his first approval after being convinced that the proposed bridge would pose no hindrance to navigation due to insufficient clearance (the 228-foot underclearance specified for the bridge easily accommodated the largest aircraft carriers in the fleet), and that if the bridge was destroyed its rubble would not choke the Narrows. In 1954, the TBTA and the Port Authority produced a joint transportation study calling for the bridge, and Moses engaged the universally respected Ammann to design the great structure. These two developments impressed lawmakers in Albany, and in 1957 the state legislature granted approval to the TBTA. While politicking with the state government, Moses petitioned the Army to obtain easements to construct tower foundations, anchorages, and approaches. After two years of tense meetings, the Department of the Army granted easements through Fort Hamilton in Brooklyn and Fort Wadsworth on Staten Island in return for $24 million in new facility construction on other Army property.

Meanwhile, efforts were being made to acquire land to build the stretches of circumferential highway to be served by the bridge. In July 1957, the New York City Board of Estimate approved the purchase of land for the Staten Island Expressway and the Third Avenue section of the Brooklyn–Queens Expressway; in March 1959, land acquisition for the Seventh Avenue section of the Brooklyn–Queens Expressway was approved. An approval from the New York City Council posed the final barrier. Before the City Council was swayed, Moses had to battle grassroots resistance and pressure brought by some of New York's most distinguished intellectuals. Seizing the controversy as an opportunity to take over the design commission from Ammann, Steinman began promoting a bridge that would bypass Bay Ridge and Staten Island altogether, linking northern Brooklyn to Bayonne, New Jersey. Running for governor in 1958, Nelson Rockefeller sided with the grassroots opposition, adding momentum to the Steinman proposal. Moses later persuaded Governor Rockefeller to reverse his position, and early in the summer of 1959, City Council capitulated.

At the time that Moses won final approval, the finances of the TBTA were stretched, and the authority could not afford to build the bridge by itself. Moses therefore persuaded the Port Authority of New York to provide start-up funds (the Port Authority was an enthusiastic backer of the bridge, if only because it promised to stimulate toll revenues on the underused New Jersey–Staten Island crossings). Construction on the Verrazano-Narrows began in August 1959. By the end of the year, the TBTA was able to take over full financial responsibility for the project. A planning vision seventy-one years in the making was about to take form.

THE FINAL WORK OF NEW YORK'S PREEMINENT BRIDGE DESIGNER

The Verrazano-Narrows Bridge, comprising a few finely tooled structural members, is a classic modernist suspension bridge. Its towers are soaring monoliths standing in clear water. The graceful catenary curve of the suspension cable supports a slender, meticulously articulated deck structure. The ageless, clean-cut lines of the anchorages bracket the path of the cable while serving as stepping-stones to the roadway as it leaps from land to air. The material reduction and linear eloquence of the bridge epitomizes the engineer's mature style. True to Ammann's description of its conception, the bridge is "an enormous object drawn as faintly as possible."[1]

The bridge is indeed enormous. It was the world's longest and heaviest suspension bridge at the time it opened. Although its clear span is only 60 feet longer than that of the Golden Gate Bridge, its suspended load is 75 percent greater. Because of its unprecedented length (4,260 feet), the curvature of the earth had to be taken into account when the towers were built: although both towers are plumb, their tops are one and five-eighths inches farther apart than their bases. Temperature has a dramatic impact on the roadway. During the swing from zero degrees Fahrenheit on a cold winter's day to a hundred-degree summer scorcher, the road deck at the center of the span rises and falls a total of twelve feet. The bridge contains enough concrete to build a single-lane highway from New York to Washington, D.C. Its cable wire, if laid end to end, would reach more than halfway from

7.3 The Verrazano-Narrows Bridge serves as a gateway between the Upper and Lower Harbors for ships and motor vehicles alike. (Photo, Jet Lowe; courtesy of the Historic American Engineering Record, National Park Service)

bottom left

7.4 Interior of the lower deck of the suspended, two-level roadway. (Photo, Jet Lowe; courtesy of the Historic American Engineering Record, National Park Service)

bottom right

7.5 A section of pre-assembled road deck truss being mechanically lifted into position and attached to the bridge's suspender cables. (Photographer unknown; courtesy of MTA Bridges and Tunnels, Special Archives)

7.6 A rocker link fastened to the steel-plate tower. Rocker links take the place of suspender cables where the road deck passes through the tower. (Photo, Dave Frieder; courtesy of Dave Frieder)

Times Square to the moon. Its towers are as high as a seventy-story building and contain as much steel as the Empire State Building. The total weight of the bridge is an astounding 1,265,000 tons. By contrast, the Empire State Building weighs a mere 365,000 tons.

The graceful appearance of the towers belies their internal complexity. The tower shafts—as well as the connecting portal struts—are of cellular construction comprising preassembled web plates interconnected by angles and erected in forty-foot lifts (figs. 7.19, 7.20). Each tower has ten thousand steel cells riveted and bolted one on top of the other and side by side; six million rivets and two million bolts hold the towers together. All the cells are numbered and are accessible from the interior of the tower, grouped around a large central shaft that accommodates elevators for convenient inspection and maintenance. Partly for aesthetic effect and largely for economy, the towers are tapered from top to bottom, growing wider as the weight of their loads increases (fig. 7.11).

The design of the bridge's long, suspended two-level roadway presented a major challenge to Ammann and his support team. Financial constraints dictated material economy, but a lightweight structure had to be balanced against the need for adequate resistance against static and dynamic forces. The design solution is two lightweight, relatively low vertical stiffening trusses (figs. 7.4, 7.5). Torsional stiffness was economically achieved by arranging lateral trusses in the plane of both the upper and lower decks. The stiffness of the steel skeleton was further enhanced by rigidly connecting all deck members: the transverse beams carrying the deck slab, the stringers of both decks, the rigid-frame floor beams, the two vertical stiffening trusses at the end of the floor frames, and the two lateral trusses per deck. In this unique design, all parts participate in resisting the vertical, torsional, and lateral forces from moving loads and wind.

CAISSON FOUNDATIONS

To optimize the length ratio between the side spans and the central span, the engineer found it necessary to site the towers offshore. The placement created no obstacle to navigation through the Narrows, but constructing foundations underwater posed a classic engineering problem.

In the nineteenth century, the French developed—and Americans perfected—an ingenious method of building underwater foundations, a form of which was followed for the Verrazano-Narrows Bridge. Known as caisson construction (*caisson* is the French word for box), its history began with enormous open-ended, watertight containers that were pushed

7.7 Profile of the road deck. The rocker links are visible on the insides of the tower, just above the road deck. (Photo, Jet Lowe; courtesy of the Historic American Engineering Record, National Park Service)

underwater to cover the area where the foundation was to be excavated and constructed. Such caissons were forced underwater with their open end down so that a work chamber with an air cavity would be in place once the caisson was embedded in soil. An airtight passage was then installed from the surface of the water to the sunken caisson. Through the passage flowed excavators, excavated soil, material for the foundation, and fresh air.

This initial form of caisson construction proved to be a problem when practiced at great depths. Workers entering and exiting the caissons often suffered a mysterious combination of ailments that occasionally resulted in death. The medical mystery of "caisson sickness" was solved during construction of the Brooklyn Bridge, but only after the second chief engineer for the bridge's construction, Washington Roebling, himself suffered a debilitating form of the illness. Caisson sickness is known more commonly today as decompression sickness or the bends: a condition brought on by too rapid a change in the air pressure surrounding the body. Once the cause was understood, a combination of decompression chambers and more carefully controlled entrances and exits eliminated the problem and made way for the development of more daring and innovative caisson techniques.

The caisson technology used for the Verrazano-Narrows Bridge bears little resemblance to nineteenth-century techniques. Workmen at the Verrazano-Narrows never entered its caissons. Furthermore, its caissons *are* the foundations for the bridge's towers. Constructing these caissons involved a simple yet ingenious process.

The caissons were built in place. An existing island where Fort Lafayette once stood served as the site for the Brooklyn tower's caisson-type foundation; a man-made island was built as the construction platform for the Staten Island caisson. Eventually the two caisson foundations came to rest 170 and 105 feet, respectively, below the surface of the water. Reaching down to that depth entailed a dual technique of construction followed by submersion.

During construction, the thirteen-foot-high base of each caisson was left open. Above this base was progressively built a gridlike arrangement of sixty-six vertical shafts—each seventeen feet in diameter—formed with reinforced concrete. Each time workers built up the height of the shafts to approximately forty feet above ground, cranes with "clamshell buckets" reached into them to dredge out the muck and sand at their base. As the earth supporting the caisson was removed, the caisson sank into the ground. To facilitate submersion, rings of water jets were placed at the exterior with nozzles pointing upward, and water released through the jets reduced friction along the outer wall during the sinking process.

When each caisson reached its predetermined depth, the base of the caisson was sealed

7.8 Construction of the foundation caisson for the Brooklyn tower. (Photo, Paul Rubenstein; courtesy of MTA Bridges and Tunnels, Special Archives)

bottom left

7.9 This aerial photograph, taken above Bay Ridge in Brooklyn, shows construction of the tower foundation caisson offshore, with initial excavation of the Brooklyn anchorage under way in the foreground. (Photo, Paul Rubenstein; courtesy of MTA Bridges and Tunnels, Special Archives)

bottom right

7.10 The Staten Island caisson after the fifth pour of concrete, with the process of sinking the pour under way. Shovels attached to cranes are removing the riverbed under the foundation and loading it onto barges. (Photo, Paul Rubenstein; courtesy of MTA Bridges and Tunnels, Special Archives)

7.11 Profile of the Staten Island tower resting on the caisson foundation. (Photo, Jet Lowe; courtesy of the Historic American Engineering Record, National Park Service)

with concrete, the vertical dredging shafts were filled with water, and the top of the caisson was capped with a reinforced concrete slab engineered to evenly distribute the weight of the tower throughout the foundation. The transition between steel tower and distribution slab is made with twenty-eight-foot concrete pier pedestals that are covered with a granite face.

THE ANCHORAGES

The cavernous anchorages of New York's suspension bridges are among the city's most breathtaking spaces. A visit to the interior is an encounter with overwhelming scale. A visitor relies more on the sense of sound than sight to gauge the boundaries of the enclosure: the acoustical qualities of the space are an eerie mixture of heavy silence crouched under a canopy of baffled echoes.

Anchorages resist the phenomenal pull of the bridge's suspension cables. Anchorage size is therefore directly proportional to the amount of force transferred. The transferral of force is a two-stage process. The first point of transferral occurs at the front of the anchorage, where the compacted cables bend around saddles resting on inclined steel posts, also known as cable bents—they are part of the steelwork visible at the front of the anchorage. From the saddle at the top of the post, the cable strands are splayed to fan out over a length of approximately one hundred feet before winding around the strand shoes at the exposed tops of the eyebars. The eyebars transfer the remaining force to inclined girders buried deep within the concrete heel at the rear of the anchorage.

The spread footings on which the anchorages were constructed are 230 feet and 345 feet long—roughly the area needed to place two football fields side by side. These footings rest on compact glacial sands at 52 and 76 feet below ground level for the Brooklyn and Staten Island anchorages, respectively. The area of excavation for the west anchorage required the movement of almost half a million cubic yards of earth. Together, the anchorages contain 780,000 tons of steel and concrete.

The visual prominence of the ten-story structures demanded careful aesthetic consideration. The triangular shape expresses the anchorage's function of resisting the pull of the inclined suspension cable. Visible grooves in the finished concrete surface relieve the monotony of the monumental concrete prism, reinforce the profile of the sculptural form, and conceal expansion joints that accommodate movement and control surface cracking. Exposed concrete was hand cleaned and polished.

The anchorages of New York's suspension bridges have been put to a number of uses, from farmer's markets to performance spaces. A proposal was once circulated to use a Brooklyn Bridge anchorage as a repository for gold bullion. Records of the planning, construction, and operation of the bridges have often been stored in their own anchorages. Unique among these structures in New York, the Staten Island anchorage of the Verrazano-Narrows Bridge contains a heating plant where cinders are warmed before being loaded

7.12 The anchorage on the Brooklyn side. (Photo, Jet Lowe; courtesy of
the Historic American Engineering Record, National Park Service)

7.14 Construction of the Staten Island anchorage, with the first section of the sixteenth tier of the tower just under way. (Photo, Paul Rubenstein; courtesy of MTA Bridges and Tunnels, Special Archives)

top

7.13 Interior of the Staten Island anchorage. (Photo, Jet Lowe; courtesy of the Historic American Engineering Record, National Park Service)

bottom

7.15 Spinning cable wire around eyebars before the road deck enclosed the anchorage. (Photo, Paul Rubenstein; courtesy of MTA Bridges and Tunnels, Special Archives)

7.16 Eyebars connect the wires of the suspension cables to the anchorage.
(Photo, Jet Lowe; courtesy of the Historic American Engineering Record, National Park
Service)

7.17 Salt and cinders are stored in the interior of the Staten Island anchorage for use
on the roadway during icy weather. (Photo, Jet Lowe; courtesy of the Historic American
Engineering Record, National Park Service)

onto trucks and spread by maintenance workers on snow and ice that accumulates on the bridge's road deck.

IRON WORKERS WITH NERVES OF STEEL

All great bridges were built by hardened, colorful men. Twelve thousand worked on the Verrazano-Narrows Bridge; one thousand men were on the site during the construction schedule's peak. Each of them was either a "boomer" or a local union member.

Boomers are itinerant builders and site supervisors who pack up at the end of each project and move on to the next, no matter the distance. Many of the boomers who came to the Verrazano-Narrows had previously worked on the Walt Whitman, Mackinac, and even the Golden Gate Bridges. Some had worked on recently erected skyscrapers in Chicago and Philadelphia. The top dog of the entire crew, John Murphy (feared and known by all subordinates as "Hard Nose" Murphy) was a boomer. He had zigzagged across the United States during a long career that included steel connecting at the George Washington Bridge in 1930–1931. Boomers possess a swagger and mystique that goads the envy and competitive spirit of the local men.

Those on the site who were not boomers were union hall regulars, most of them members of either Brooklyn Local 361 or Manhattan Local 40. They had their own distinguished history of heroic construction: their members had had a hand in all the landmark bridges built in New York, as well as the city's majestic skyscrapers.

Among iron workers, be they boomers or hometown boys, the operative hierarchy governing the site was a division of labor that marked each boomer or union regular as one of the following: a punk, a full-fledged bridgeman, a pusher, a walkin' boss, or *the* boss. Punks are apprentices who fetch, observe, and occasionally get a turn at the tools. After two or three years, punks graduate to become regular bridgemen. Bridgemen are the men who heat, catch, and drive rivets and position, connect, and weld steel. Apprentices and bridgemen are closely supervised by pushers: Pushers are entry-level managers responsible for driving the daily productivity of small work gangs. Above the pushers are walkin' bosses who assist the superintendent of construction. There were four walkin' bosses for the Verrazano-Narrows Bridge, each in charge of all work on one of four sections of the structure. They—and every worker at the site—answered to Hard Nose Murphy, the superintendent of construction (*the* boss, the top dog), and he answered to both the general manager, George E. Spargo (who was succeeded by Peter J. Reidy), and the chief engineer, Ammann.

7.18 A cable-spinning crew at the summit of the Brooklyn tower. (Photo, Paul Rubenstein/Lenox Studios Photography; courtesy of MTA Bridges and Tunnels, Special Archives)

7.19 Preassembled sections of the towers are lowered into position and bolted to adjacent sections. (Photographer unknown; courtesy of MTA Bridges and Tunnels, Special Archives)

7.20 Barrels and buckets of bolts and rivets were scattered throughout the towers during the assembly process. (Photo, Paul Rubenstein/Lenox Studios Photography; courtesy of MTA Bridges and Tunnels, Special Archives)

7.21 Electricians install utility lines along the temporary walkway. Cable spinning has not yet begun. (Photographer unknown; courtesy of MTA Bridges and Tunnels, Special Archives)

7.22 A suspender cable saddle being lowered into position. Men operating a movable platform along the suspension cable, partially visible at upper left, are compacting and banding the cable. (Photographer unknown; courtesy of MTA Bridges and Tunnels, Special Archives)

7.23 Iron workers prepare to attach a saddle to one of the main suspension cables. Once the saddle is clamped, a suspender cable will be attached to it. (Photographer unknown; courtesy of MTA Bridges and Tunnels, Special Archives)

To say that iron workers have nerves of steel is no understatement. Serious risk to limb and life faces each person who enters the construction site. The tragedies and near tragedies witnessed by the older men are many and often legendary. Tragedy struck the Verrazano-Narrows, where three men fell to their deaths during construction. One fell from a ladder inside a tower; a second slipped while working on an elevated approach road; and a third plunged from a catwalk during the cable-spinning operation.

Gerard McKee—a nineteen-year-old iron worker who fell to his death from a temporary walkway during cable spinning—was a much-admired young Brooklyn man whose bridge-building father had proudly introduced him and his two older brothers to the union hall. It remains unclear how McKee lost his balance on the morning of October 9, 1963. He was hanging from the walkway by one hand when his partner, Edward Iannielli, heard him cry out for help. By the time he reached him, the much smaller and lighter Iannielli—whose left hand had been disabled two years earlier when caught in a crane—was unable to pull his partner to safety. After a valiant struggle, a silenced, stunned workforce watched McKee's long fall. As he fell, his shirt was ripped off by the wind. From the bridge the outline of his pale back receded against the dark waters of the East River as he dropped to his death. No one moved. A few began to weep. All removed their hard hats and quietly left the site.[2]

In the wake of McKee's death, union leaders demanded higher safety standards, arguing that the three deaths that had occurred on the bridge could have been avoided had there been construction nets. Management stalled but eventually agreed to hang safety netting after workers walked off the job in protest. Before construction was completed, four workmen fell and were caught by the nets.

Too often, the builders' role in a bridge's magic is left unsung, and such was the case with the Verrazano-Narrows Bridge. None of the workers were sent invitations to the opening ceremony by the event's organizers, so none joined the crowds that thronged the podium where VIPs assembled to celebrate the flawless placement of every wire, rivet, beam, and cubic yard of concrete in this amazing work of aerial architecture.

NEIGHBORHOOD PRESERVATION VERSUS REGIONAL DEVELOPMENT

The western expressway to the Verrazano-Narrows Bridge was paved across the hinterlands of New York's most rural borough, terminating in a generous one-level toll plaza. By contrast, the enormous traffic interchange at the bridge's eastern approach gutted a close-knit, densely populated Brooklyn community—Bay Ridge—before cutting a wide swath northward. Many New Yorkers considered the plan an abomination because it cut so deeply into long-established neighborhoods. Public interest groups actively protested during the legislative approval process. Most vocal among them was the "Save Bay Ridge" movement. Its eight thousand members were directly affected by the twelve-lane approach, which they dubbed the "Colossus of Roads."

A number of prominent New York intellectuals joined the opposition movement. The leading architecture and planning critic of the day, Lewis Mumford, characterized the pro-

7.24 The bridge's toll plaza, at far left, is on the Staten Island side of the suspension structure. Both levels of bridge traffic share a single wide plaza built to accommodate twenty-eight toll collection booths. (Photo, Jet Lowe; courtesy of the Historic American Engineering Record, National Park Service)

posal as "a menacing error . . . a terrifying project." In a critique published in the "Skyline" column of the *New Yorker* on November 14, 1959, Mumford went on to say, "At the very moment . . . that we have torn down our elevated railways, because of their spoilage of urban space, our highway engineers are using vast sums of public money to restore the same nuisance in an even noisier and more insistent form. But what is Brooklyn to the highway engineer—except a place to go through quickly, at whatever necessary sacrifice of peace and amenity by its inhabitants?"[3] But other respected voices endorsed the project, including the editorial board of the *New York Times*. Ultimately, Bay Ridge would have a bridge.

The regional impact of the highway exceeded the rosiest expectations of its planners and engineers. Taking advantage of the new artery for commercial trade to the city's great air, sea, and rail ports, commerce flourished. New industries located near the highway, creating jobs, and with them, booming residential communities. As testimony to the surge precipitated by the bridge, its lower deck opened to traffic in 1969—more than six years ahead of anticipated need.

The benefits of the highway to regional development have never been disputed. The local impact of the highway continues to prod debate. To make way for the bridge's approaches, eight hundred buildings were demolished and seven thousand residents displaced in Bay Ridge alone. On less densely populated Staten Island, four hundred buildings were lost and thirty-five hundred islanders uprooted. Many of the affected residents on both sides of the Narrows were private homeowners, elderly and deeply shaken when ordered to abandon households that had witnessed family marriages, births, and deaths.

As was expected, the community cohesion enjoyed by the "prebridge" Bay Ridge residents was decisively and immediately shattered by the broad ribbons of concrete roads and viaducts that physically severed the neighborhood. On Staten Island, the pace of change accompanying the opening of the expressway was comparatively slow. Dirt roads and strawberry fields surrounded communities along the Clove Lakes Expressway well into the mid-1970s. But by the mid-1990s the Greek- and Italian-Americans who had lived in the same Staten Island communities since before the bridge's construction—residents who had once derived their living from farming—lamented that they had witnessed their small-town life gradually subsumed by population increases, commercial development, and the general displacement of agriculture with industry.[4] None of these changes had come from within. All were based on regional trends and regional markets.

Change is inevitable, but it is also manageable. A prime challenge that confronted the planners of the Verrazano-Narrows Bridge—whether they competently addressed it or

7.25 An aerial view of the bridge showing the Brooklyn interchange at lower right and the connecting highway on Staten Island cutting a regional-scale path across the largely rural landscape. This photograph was taken shortly before the bridge opened. (Photographer unknown; courtesy of MTA Bridges and Tunnels, Special Archives)

bottom left

7.26 The east tower from a street in Bay Ridge, Brooklyn. (Photo, Jet Lowe; courtesy of the Historic American Engineering Record, National Park Service)

bottom right

7.27 The traffic interchange on the Brooklyn side is made with long, narrow loops. This minimized the amount of neighborhood demolition, but the bridge forever changed the structure of the community all the same. (Photographer unknown; courtesy of MTA Bridges and Tunnels, Special Archives)

7.28 Construction of the Brooklyn–Queens Expressway (Interstate Route 278) leading to the bridge was preceded by the demolition of hundreds of homes and commercial facilities in the close-knit neighborhood of Bay Ridge. (Photographer unknown; courtesy of MTA Bridges and Tunnels, Special Archives)

not—was to expand regional opportunities while preserving local constituencies that are vital and healthy. This challenge is best met by an enlightened process of consensus leading to a workable formula for sustainable growth. Based on the experience of the Verrazano-Narrows Bridge and highway project, it seems fair to say that meeting the challenge of community preservation proved more difficult than engineering the world's longest suspension bridge.

Ammann was present at the opening of the bridge on November 21, 1964, when Robert Moses, as chairman of the Triborough Bridge and Tunnel Authority, hailed the structure as "a triumph of simplicity, and of restraint over exuberance."[5] The engineer died the following September at eighty-six. Among the designers of New York's long-span bridges, Ammann has no rivals, and among Ammann's New York bridges, the Verrazano-Narrows stands unmatched in the purity of its architectural form.

Appendix Ammann's Built and Unbuilt Projects

In addition to Othmar H. Ammann's role as the guiding creative force and principal author of the six bridges featured in this volume, he was the principal designer and chief engineer of two smaller structures built in New York City: the Little Hell Gate Bridge and the Harlem River Pedestrian Bridge.

At the peak of his career, Ammann designed several projects that were not built. Ample documentation for four of these structures exists so that attribution can be confidently assigned. Information concerning these unbuilt projects, as well as those that were built, follows.

Crediting Ammann's involvement with a number of other projects to which he lent a hand is more difficult. Of the numerous bridges he worked on in a supporting capacity during his early career, the Hell Gate Bridge (1917) stands out because of Ammann's level of responsibility as chief assistant to Gustav Lindenthal. After his independent mastery was established with the design of the George Washington Bridge, Ammann was routinely asked to serve as a design consultant on the long-span projects of other civil engineers. Among those projects were the Golden Gate Bridge in San Francisco (1937), the Delaware Memorial Bridge in Wilmington, Delaware (1951), the Walt Whitman Bridge over the Delaware River in Philadelphia (1957), and the Mackinac Bridge in Mackinaw, Michigan (1958). Ammann's involvement with the Golden Gate is the most intriguing of the four. Joseph B. Strauss was awarded the commission in 1929 for a bridge scheme he had been promoting since 1925. His initial design was an expensive and ungainly truss bridge. The members of the Golden Gate Bridge Commission who had hired Strauss were appalled by the design but also reluctant to directly request changes. Instead, the commission put together an advisory panel of engineers to help Strauss with a redesign. Ammann was named to the panel but considered it imprudent to play a prominent public role because he was a full-time public servant to the states of New York and New Jersey as chief engineer for the Port Authority. He was, however, very active in the Golden Gate design, filtering his recommendations through a Port Authority colleague, Allston Dana. Ammann's influence on the design of the Golden Gate can be easily inferred in the completed bridge—an elegant suspension structure with stiffening trusses beneath its road deck.

GEORGE WASHINGTON BRIDGE

Connecting Fort Lee, New Jersey, and West 178th Street in Manhattan

Othmar H. Ammann, Chief Engineer
Edward W. Stearns, Assistant Engineer
Leon S. Moisseiff, Advisory Engineer of Design
Allston Dana, Engineer of Design
Montgomery B. Case, Engineer of Construction
Robert A. Lesher, Traffic Engineer
Prof. William H. Burr, Consulting Engineer
Gen. George W. Goethals, Consulting Engineer
Daniel E. Moran, Consultant on Foundations
Cass Gilbert, Consulting Architect

Design inception: 1923
Commencement of construction: October 21, 1927
Opening date: October 25, 1931
Commencement of construction of the lower six-lane road deck: September 1958
Opening date of lower deck: August 29, 1962
Owned and operated by the Port Authority of New York and New Jersey

Originally known as the Hudson River Bridge, the George Washington Bridge has a 3,500-foot clear span, more than double that of any suspension bridge previously built. Engineering the span was not merely an exercise in pushing the boundaries of established practice—a new conception regarding the behavior of long-span structures was necessarily put into play. This, as much as its record-breaking length, establishes the George Washington's prominence in the history of bridge design.

The bridge is subtly asymmetrical: the foundation of the New York tower was built on the riverbank at Fort Washington Point; the foundation for the New Jersey tower is built beneath the Hudson River approximately seventy-six feet from shore at a place where bedrock lies less than one hundred feet beneath the river. Sites for the anchorages are also asymmetrical: the New York anchorage is a massive concrete structure resting on the upper bank of the river; on the New Jersey side, the main suspension cables are anchored in the bedrock of the Palisades.

The George Washington's towers are built of rigidly connected three-dimensional steel frames with arched portal bracing. The steel was designed to support itself, the suspension cables, and the road deck during construction, and the towers were to be encased with concrete as reinforcement before the bridge opened to the live loads imposed by vehicular traffic. The bridge was a beneficiary of an innovative American steel industry constantly producing improved material, and once it became clear that the steel alone would be sufficient to support the structure, the supervisors eliminated the concrete. Ammann and consulting architect Cass Gilbert argued on aesthetic grounds for preserving the architectural integrity of the original design by wrapping the steel in ornamental stonework, but their arguments were not as persuasive as the impetus to cut costs. Contracts for supplying and erecting the stone curtain walls were never awarded, and the steel structure of the towers remains exposed.

The bridge's four main suspension cables are each wound from 61 strands with each strand comprising 434 wires. After being compacted and wrapped, the cables are just under three feet in diameter. The suspender ropes from which the floor system is hung are composed of galvanized wires arranged into six strands around an independent wire rope center. On both sides, the main cable splays at the anchorages. The strands fan out and are attached to steel eyebars connected to anchoring girders. The eyebars and steel girders are all encased in solid concrete.

The upper-deck roadway consists of a built-up plate-girder floor beam, attached to the wire suspender ropes hanging from the main cables. Into the floor beams are framed the plate-girder stringers that carry the cross beams. On top of the cross beams are bulb beams embedded in the concrete floor slab. The stability of the bridge against torsion was enhanced with the addition of the lower deck in 1962.

SPECIFICATION HIGHLIGHTS
- 14 traffic lanes: 8 on upper level, 6 on lower level (original structure consisted of a single deck with six lanes, three on each side of an open area, which was later used for two more traffic lanes added in 1946)
- Length of clear span: 3,500 feet
- Length of span between anchorages: 4,760 feet
- Width of bridge: 119 feet

- Width of roadway: 90 feet
- Height of towers above water: 604 feet
- Channel clearance of bridge at midspan: 212 feet
- Towers incorporate 40,000 tons of structural steel. The tower is constructed with prefabricated sections each weighing 37 to 80 tons. Each tower has 475,000 rivets weighing a total of 325 tons.
- Each tower has two elevators: one that operates by rack and pinion and was installed during initial construction, and one run by a cable that was added later. (Designs for a towertop restaurant and viewing deck were considered but never implemented.)
- There are 107,000 miles of wire in the main suspension cables (if stretched out, the wire would reach nearly halfway to the moon)
- New York anchorage used approximately 110,000 cubic yards of concrete and weighs 350,000 tons
- New York anchorage contains 1,400 eyebars; New Jersey anchorage contains 2,100 eyebars. Eyebars are approximately 40 feet long and weigh more than a ton each.
- Design provides for a maximum lateral swing of 5 feet

BAYONNE BRIDGE

Connecting Bayonne, New Jersey, and Port Richmond, New York

Othmar H. Ammann, Chief Engineer
Edward W. Stearns, Chief Assistant
John C. Evans, Terminal Engineer
Allston Dana, Engineer of Design
Montgomery B. Case and W. J. Boucher, Engineers of Construction
Leon S. Moisseiff, Consulting Engineer for Design
Cass Gilbert, Consulting Architect

Design inception: 1926
Commencement of construction: September 1928
Opening date: November 14, 1931
Owned and operated by the Port Authority of New York and New Jersey

The Bayonne Bridge's breathtaking span is carried by a two-hinged spandrel-braced trussed arch in which the bottom chord carries most of the dead load

A.1 Ammann delivering an address at the opening ceremony for the Bayonne Bridge, November 14, 1931. (Photographer unknown; courtesy of Margot Ammann-Durrer)

and uniform live load, and the top chord and web members are stressed principally by live loads and temperature. A hinge at the springing of the arch consists of a sixteen-inch-diameter pin threaded through a steel forging, which in turn rests on riveted structural-steel arch shoes cast into the skewbacks.

The suspended roadway is hung from the trussed arch with wire rope hangers. Floor beams and stringers supporting the concrete deck slabs are of silicon and carbon steel. The abutments, founded on bedrock, are built of solid concrete. Granite facing covers part of the concrete from just beneath the ground line to the height of the the arch shoes. The abutments were designed to have a granite-clad superstructure rising to the level of the road deck; the steel framework for the granite was partially erected, but the stone was never hung. Ornamentally significant, the pseudo-abutment appears to be resisting the thrust of the arch, making a formal transition between the viaduct and the arch span.

The long viaduct approaches to the arch consist of deck girder spans resting on piers of concrete reinforced with structural steel frames.

Until recently, the Bayonne Bridge boasted the longest arch span in the world, exceeding the Sydney Harbor Bridge in Australia by twenty-five feet. The New River Gorge Bridge in Fayetteville, West Virginia, currently is the longest arch at 1,699 1/2 feet.

SPECIFICATION HIGHLIGHTS

- Bridge and approaches contain 26,000 tons of steel (17,000 tons in the span alone)
- Bridge is positioned at a 58-degree angle to the shoreline
- Clear span made by arch: 1,675 feet
- The Port Richmond viaduct approach is approximately 2,900 feet in length and the Bayonne viaduct approach about 3,700 feet, making the total length of the structure from plaza to plaza roughly 1 2/3 miles long
- The arch delivers a 29.4-ton thrust to each abutment
- Width between arch trusses: 74 feet
- Clearance between the mean water height and bottom of road deck: 150 feet

TRIBOROUGH BRIDGE

Connecting the boroughs of Manhattan, the Bronx, and Queens around Randall's Island, New York City

Othmar H. Ammann, Chief Engineer
Edward W. Stearns, Assistant Chief Engineer
Allston Dana, Engineer of Design
John C. Evans, Engineer of Approaches
Col. H. W. Hudson, Engineer of Construction
Aymar Embury II, Architect
Ash-Howard-Needles and Tammen, Consulting Engineers (truss bridge)
Leon S. Moisseiff, Consulting Engineer
Daniel E. Moran, Consulting Engineer on foundations

Design inception: 1916
Completion of feasibility study: 1925
Commencement of construction: October 25, 1929
Project discontinued due to lack of funds: 1932
Project transferred and reactivated by the Triborough Bridge Authority: 1933

Opening date: July 11, 1936
Owned and operated by the Triborough Bridge and Tunnel Authority

The Triborough Bridge complex is a three-and-a-half-mile, three-branched waterway crossing, comprising three long-span bridges and an extensive viaduct network connecting three New York City boroughs. The complex includes

- the Triborough Suspension Bridge, crossing from Randall's Island to Astoria in Queens
- the Triborough Lift Bridge, connecting Manhattan to Randall's Island
- the Triborough Truss Bridge (designed to be convertible to a lift bridge), crossing the Bronx Kill between Randall's Island and the Bronx
- two and a half miles of viaduct
- an innovative three-legged toll-traffic exchange on Randall's Island joining the lift, truss, and suspension bridges and handling twelve different traffic movements on two separate toll plazas.
- approximately fifteen miles of arterial highway approaches

Of the four distinct bridge forms comprising the Triborough Bridge, Ammann designed three: the suspension bridge, the lift bridge, and the extensive viaduct. The Triborough Truss Bridge was designed by Ash-Howard-Needles and Tammen, Consulting Engineers.

The Triborough Suspension Bridge is often mistakenly thought to be the entire Triborough Bridge. Its two main cables were spun between two cruciform-shaped towers. Each tower consists of two cellular columns—built of silicon steel plates and angles—connected by cross-braced panel trusses of carbon steel. The towers are trussed at three different levels—just below the roadway, just above the roadway, and at the very top of the tower structures. Cast steel cable saddles in the top sections of the towers support galvanized steel suspension cables. Steel suspender ropes are attached to the main cables to carry the roadway. In the roadways, silicon steel plate girders support rolled stringers that in turn carry the steel cross beams and concrete roadway. Warren trusses stabilize the deck structure.

The Triborough Lift Bridge has a movable roadway that can be raised to allow tall boats to pass under it. There are five main parts to the bridge: two steel towers and three truss segments. The middle truss can be vertically lifted. Electric motors housed at the top of the towers hoist the bridge's 2,050-ton central road span 135 feet above the water level of the Harlem River. The other two trusses span from the approach road to the movable deck. All three trusses are independent structures.

The 13,500 feet of viaducts connecting the major spans are constructed with three lines of plate girders resting on concrete piers. Floor beams are framed between the girders, as are I-beam stringers that support steel cross beams and the concrete roadway slab.

SPECIFICATION HIGHLIGHTS (TRIBOROUGH SUSPENSION BRIDGE)
- Deck carries 8 lanes of traffic
- Clear span between towers: 1,380 feet
- Tower height: 290 feet 8 inches from concrete pier to base of cable saddle
- Towers comprise 5,500 tons of steel. Silicon steel was used for the legs and carbon steel for bracing members.
- Towers have a maximum deflection of 14 inches under worst possible conditions of combined unbalanced live loads and temperature
- Floor system is designed for a dead load averaging 19,200 pounds per linear foot, live load of 4,000 pounds per linear foot, and wind load of 1,200 pounds per linear foot
- Floor beams are plate girders 96 feet long and 8 feet 4 1/2 inches deep fabricated with silicon steel
- Each suspension cable contains 37 strands of 248 cold-drawn galvanized steel wires (each wire is 0.196 inches in diameter)
- Stiffening truss in the road deck is a Warren truss web system that is 20 feet deep and hinged at the tower connections
- Anchorages are 150 feet by 225 feet and are founded on bedrock. The Wards Island anchorage weighs 115,000 tons and contains 59,000 cubic yards of concrete; the Queens anchorage contains 74,500 cubic yards of concrete.

LITTLE HELL GATE BRIDGE

Connecting Randall's and Wards Islands, New York City

Othmar H. Ammann, Chief Engineer

Edward W. Stearns, Assistant Chief Engineer

Allston Dana, Engineer of Design

Leon S. Moisseiff, Consulting Engineer

Opening date: July 11, 1936

Demolition: summer 1996

Although not remarkable for its length, the Little Hell Gate Bridge was a jewel among Ammann's works, linking parks and institutions built on Randall's and Wards Islands.

The bridge stood between the Triborough Bridge's viaduct and the approach viaduct to the Hell Gate Railroad Bridge. Because the deck is considerably lower than those of the larger structures, travelers crossing the bigger bridges had no view of the Little Hell Gate, and only visitors to Randall's and Wards Islands saw this superb structure.

The bridge included reinforced concrete approach viaducts leading to

A.3 Aerial view, looking south, of the Triborough Bridge complex, showing the Little Hell Gate Bridge between the Hell Gate Bridge, at left, and the Triborough Viaduct. (Photographer unknown; courtesy of MTA Bridges and Tunnels, Special Archives)

three arched spans: the center span of 265 feet 6 inches; from Wards Island to the center span, 236 feet 3 inches; and from Randall's Island to the center span, 161 feet. Concrete abutments separated the arches. The arches were of girder-plate construction with steel cross bracing between arch ribs. Steel spandrel posts rested on the arches to carry the road deck above the profile of the arch. The deck was 24 feet wide, and pedestrian walkways 7 feet wide were cantilevered beyond it.

The channel separating Wards and Randall's Islands was fully filled in by 1980, nudging the bridge toward obsolescence. A decline in use of the islands' deteriorating recreational facilities, and the transfer of the bridge's ownership from the Triborough Bridge and Tunnel Authority to the New York State Department of Transportation, did not augur well for the structure. Arguing

A.2 The Little Hell Gate Bridge. (Photographer unknown; courtesy of MTA Bridges and Tunnels, Special Archives)

that maintenance costs outstripped user benefit, the Department of Transportation demolished the bridge in 1996 despite a preservation plan that would have revitalized the bridge's role as part of a well-organized park redevelopment master plan prepared by the New York City Parks Department.

This is the only bridge of Ammann's that has been torn down.

LINCOLN TUNNEL

Connecting Weehawken, New Jersey, and West Thirty-ninth Street in Manhattan

Othmar H. Ammann, Chief Engineer
Engineering Department of the Port of New York Authority, Engineers of record
Cass Gilbert, Consulting Architect

A.4 Aerial view of the traffic plaza leading to the New Jersey entrance of the Lincoln Tunnel. (Photo, Jet Lowe; courtesy of the Historic American Engineering Record, National Park Service)

Design inception: 1931
Commencement of construction: May 1934
Opening dates: (first tube) 1937; (second tube) 1945; (third tube) 1957
Owned and operated by the Port Authority of New York and New Jersey

Shortly after the Holland Tunnel's opening in 1927—and while the George Washington Bridge was still under construction—it became evident that a third highway linking New Jersey and New York City across the Hudson River would soon become necessary. The site considered ideal for it was three miles north of the Holland Tunnel and seven miles south of the George Washington Bridge. This location would accommodate escalating traffic flow in several ways: it would shorten the time and distance for westbound vehicles that originated in midtown and had to travel the many miles either south or north to reach the existing Hudson crossings; and it would align with the Queens–Midtown Tunnel under the East River to forge an east–west axis from New Jersey to Long Island. (To further strengthen the axis, the city planned a tunnel under Forty-second Street in Manhattan that would connect the Lincoln and Queens–Midtown Tunnels, but it was never constructed.)

On the New Jersey side, the tunnel location would give easy access to midtown Manhattan from important residential and industrial centers such as Newark, the Oranges, Passaic, and Paterson. The site also afforded easy connections with interstate thoroughfares and New Jersey Routes 1 and 3.

The townspeople of Weehawken demanded direct access to the Lincoln Tunnel, as did other nearby residents. Design investigations undertaken to address community concerns led to the development of a radically different approach to plaza design. Instead of driving the roadway due west through the Palisades—which would have cost less—the designer had the tunnel make a ninety-degree swing southward upon reaching the New Jersey riverbank to meet a funnel-shaped depressed plaza that connects with surface streets carrying tunnel traffic only.

The tubes of the Lincoln Tunnel are circular in cross section, thirty-one feet in diameter, cut horizontally on the interior by floor and ceiling slabs. They are constructed of twenty-three hundred cast-iron rings, each weighing twenty-one and a half tons, fastened together to form a long waterproof tube beneath the river silt. Cast-steel reinforcing strengthens areas where higher stresses form. Each ring is assembled from fourteen curved segments bolted together with high-tensile-strength bolts. The first tube runs between a ventilation shaft under King's Bluff in New Jersey and the intersection of Eleventh Avenue and Thirty-ninth Street in Manhattan.

The mile-and-a-half-long tunnel was dug by means of the "shield method," a technique that utilizes hydraulically powered steel cutting disks to push through the silt. A shield was started at each bank, meeting under the river at a point closer to the Manhattan shore. Hydraulic jacks moved the shields by pushing off the last assembled ring. Most of the material displaced by the shields was redistributed within the tunnel to keep the weight of the excavated site stable until the tunneling was complete and the air pressure equalized.

After each ring was assembled, it was bolted to the one preceding. Taper rings were used where there were changes in grade or difficult alignments. The two end segments of the tunnel, which cut through the rock of the shoreline and are constructed in a manner similar to the main section, utilize structural steel instead of cast steel or iron.

Concrete and steel slabs make up the horizontal floor and ceiling of the tube, resulting in a clear roadway height of roughly thirteen and a half feet. The leftover spaces at top and bottom serve as ventilation passageways for the tunnel: fresh-air ducts are housed beneath the roadway and an exhaust system above the ceiling.

To facilitate maintenance, instead of exposed concrete the tunnel interior is covered in a combination of glazed terra cotta and cream-colored vitreous ceramic tile set in mortar and held in place by metallic grippers.

At the entrance level of the tunnel, the New Jersey retaining walls are exposed reinforced concrete, and on the New York side the retaining walls are faced in brick and granite. The three tunnel ventilation buildings—designed in a "stepped-back" Art Deco style—are faced with large brick panels. Two of these ventilation buildings were built on conventional foundations and the third on concrete-filled steel caissons.

Work on the second tube, just north of the original passage, began as soon as the first was completed. These tubes, and a third that came later, share a common design and were constructed in an identical manner. The original traffic plazas and approach roads are also shared by all three.

SPECIFICATION HIGHLIGHTS
- Length of tunnel: 1.5 miles. All tubes have an under-river length of 4,600 feet.
- 250,000 cubic yards of rock, earth, and silt were displaced in all

- It took construction workers (called "sandhogs") 14 1/2 months to dig the first tube
- 1,300 Works Progress Administration workers were employed in the construction of the 1937 and 1945 tubes
- Top of tunnel lies 75 feet below mean high water of the Hudson River
- Distance between tubes: 75 feet center-to-center
- Roadways in each tube are 21 feet 6 inches wide
- 8,000 panes of cream-colored glass line the top of each tube and form the largest glass ceiling in the world
- A total of 26 blowers and 30 exhaust fans have a combined capacity of 8,834,000 cubic feet of air per minute to ventilate the tunnels

BRONX–WHITESTONE BRIDGE
Connecting the boroughs of the Bronx and Queens over the East River

Othmar H. Ammann, Chief Engineer
Edward W. Stearns, Assistant Chief Engineer
Allston Dana, Engineer of Design
Aymar Embury II, Architect
Leon S. Moisseiff, Consulting Engineer for Design

Design inception: 1935
Commencement of construction: June 1937
Opening date: April 29, 1939
Installation of additional stiffening and bracing: 1940, 1941, and 1946
Owned and operated by the Triborough Bridge and Tunnel Authority

The Bronx–Whitestone Bridge was the first structure of such great suspension span to apply a flexible road-deck stiffening system in the form of simple, solid plate girders. It was also the first bridge whose towers are rigid frames without any diagonal cross bracing. The simplicity and economy of the design are further reflected in the anchorages and adjacent approach viaducts, which are distilled to a minimum of materials required for strength and stability and are devoid of extraneous architectural embellishments. The external shape of the exposed concrete anchorage roughly follows the curve of the cables at their point of attachment.

The steel towers supporting the two main suspension cables are set on re-inforced concrete piers that rest on solid rock. Each tower is built of two columns of closed box-section construction connected at the top and just below the road deck by arched portal struts. The cables rest on a smooth saddle at the top of the towers. A housing was built above the saddles to protect the cable bearings.

The roadway is hung from the main cable by suspenders made of two separate ropes looped over a grooved band wrapped around the cable. The four rope ends are secured to the stiffening girders of the floor system. The main floor beams are framed into the stiffening girder and support the stringers, cross beams, and concrete slabs of the floor deck.

The approach viaducts to the bridge consist of continuous plate-girder spans, carried on reinforced concrete piers.

The structure was slightly altered in 1940 and 1941, when diagonal bracing cables connecting the roadway to the towers were added, and significantly retrofitted in 1946. Revelations regarding the potentially destructive force of wind on flexible long-span, lightweight structures precipitated a reevaluation of the deck's stability following the Tacoma Narrows Bridge failure in late 1940. Ammann concluded that the bridge was sound, but the Triborough Bridge and Tunnel Authority asked him to alter the structure for greater rigidity. As a result, fourteen-foot-tall Warren trusses were added over the plate girders. At the same time, the bridge's sidewalks were eliminated and the roadway lanes widened.

SPECIFICATION HIGHLIGHTS

- The center span is 2,300 feet and the side spans are 735 feet; total span length is 3,770 feet
- The road deck is 135 feet above mean high water at the channel line, near the Bronx shore, and approximately 150 feet at the center of the span
- Each of the main cables has a finished diameter of approximately 22 inches and a net sectional area of wire of 297 square inches
- Each main cable is fabricated with 37 strands; each strand contains 266 wires
- Anchorage dimensions: 110 feet wide by 180 feet long by 110 feet high
- The towers are 377 feet from mean high water to the tops of the cable housings
- Bridge and approaches contain 22,300 tons of steel work (exclusive of reinforcement in concrete)

A.5 Drawing of the proposed Canimar River Bridge. (Rendered by M. Michele; photo courtesy of Ann Foker)

- Approximately 200,000 cubic yards of concrete went into foundations, piers, and anchorages
- 50,000 one-inch field rivets were driven into each tower
- Bronx and Queens approach highways have a combined length of nearly 3 1/2 miles
- Centers of the two cables are 74 feet apart
- Towers have a uniform width of 99 feet from their base structures to their tops

CANIMAR RIVER BRIDGE

Connecting the coastal highway over the Canimar River near Matanzas, Cuba (approximately fifty miles east of Havana)

Othmar H. Ammann, Chief Engineer
Date of design: 1946
Designed for the National Highway Department, Cuba

In 1946, Ammann proposed engineering a six-hundred-foot span across the Canimar River with an aluminum plate-girder arch supporting spandrel posts and a road deck. It was the first long-span bridge to be designed in alu-

A.6 The Harlem River Pedestrian Bridge with the Manhattan access ramp in the foreground. The central span is in the down position. (Photographer unknown; courtesy of MTA Bridges and Tunnels, Special Archives)

A.7 The Harlem River Pedestrian Bridge, showing the central span in the "lift" position, during the final phase of the bridge's construction. (Photographer unknown; courtesy of MTA Bridges and Tunnels, Special Archives)

minum. Though considerably lighter than a steel bridge, studies showed that aluminum would also be considerably more expensive. The extra cost was, however, justifiable when compared with the savings in maintenance that rust-resistant aluminum offered over steel in the steamy, salty marine environment of Matanzas.

Part of the structure's aesthetic success is due to the complete separation of the arch from the horizontal deck girders (a similar configuration was given to the Little Hell Gate Bridge in New York). Plans for the Canimar River Bridge were discontinued in 1959 when a government headed by Fidel Castro supplanted Cuba's existing political system.

HARLEM RIVER PEDESTRIAN BRIDGE

Connecting Wards Island and East 103d Street in Manhattan

Othmar H. Ammann, Chief Engineer
Ammann & Whitney, Consulting Engineers, Engineers of record

Date of design: 1939
Commencement of construction: 1949
Opening date: 1949

Ammann retired from both the Port Authority of New York and New Jersey and the Triborough Bridge and Tunnel Authority in 1939 to establish his own consulting practice. Later that year the TBTA awarded him the commission to design a foot bridge, which he completed within months. But building material was not immediately available—most of the country's steel production was being diverted for defense purposes in connection with World War II—and construction was ultimately postponed for ten years. The office of Ammann & Whitney, Consulting Engineers, eventually oversaw construction under Ammann's direction.

Ammann affectionately referred to the structure as "the Little Green Bridge," confiding to family and friends that it was one of his favorite works. The 330-foot central span can be lifted by means of counterweights, housed within the towers, to permit passage of taller vessels.

The deck is supported by a pair of vertically ribbed plate girders twelve

A.8 Photomontage illustrating the proposed Brooklyn–Battery Bridge. (Courtesy of MTA Bridges and Tunnels, Special Archives)

feet apart. On the Wards Island side, the walkway makes a simple descent from the side span to ground level. On the Manhattan side, a system of ramps connects pedestrians to sidewalks on both sides of East River Drive by means of a viaduct across the highway.

The bridge was built to carry Manhattan residents to open park land and recreational facilities on Wards Island, but foot traffic on the bridge fell far short of estimates. That, combined with public safety concerns, has caused officials to close the bridge.

SPECIFICATION HIGHLIGHTS
· Central (operable) span: 330 feet
· Manhattan side span: 252 feet
· Combined length of Wards Island's double side span: 355 feet

BROOKLYN–BATTERY BRIDGE

Connecting the lower end of Manhattan with Brooklyn Heights in Brooklyn

Othmar H. Ammann, Chief Engineer
Proposal date: 1939
Designed for the Triborough Bridge Authority

In 1938, after much political wrangling, an agreement was reached for a bold new public project: building a tunnel to connect the southern tip of Manhattan to Brooklyn Heights. The plan was financially feasible only because Robert Moses, chairman of the Triborough Bridge Authority, stepped in and agreed to use funds from the authority's treasury to supplement the city's federal loan. For reasons that are unclear, however, in 1939 Moses reversed his position on the tunnel, refusing to provide any financial support, and instead began vigorously campaigning for a bridge between Battery Park in Manhattan and Hamilton Avenue in Brooklyn.

A bridge had several advantages over a tunnel: it could be built much faster and for less money, it would be cheaper to maintain and operate, and it could carry more traffic. Nevertheless, disadvantages related to the proposed bridge's impact on the affected neighborhoods proved more compelling, and President Franklin D. Roosevelt broke ground in 1940 for a tunnel, not a bridge. The Brooklyn–Battery Tunnel was opened to traffic in 1950.

Ammann was chief engineer for the Triborough Tunnel Authority when Moses inserted himself in the planning process for the crossing, asking Ammann to prepare the bridge plan he would champion. Ammann designed a bridge comprising two suspension structures placed end to end and sharing an anchorage pier off Governors Island. The composite structure covered a span of sixty-five hundred feet.

The proposal set off a storm of protest—though not over the visual impact of the suspension structures. Vehement criticism was directed at an elevated highway that would connect the West Side Highway to the bridge's Manhattan on-ramp along the northern boundary of Battery Park. The elevated highway, terminating at a ten-story concrete anchorage, would have degraded the pedestrian life of lower Wall Street, undermined Battery Park's charm, and irrevocably marred the New York skyline. Unrelenting opposition ultimately caused the project's demise.

Ammann argued publicly in favor of the bridge and highway proposal, but there is no evidence that he regretted the project's scuttling. Although he gave his client the best bridge for the site, one can only speculate regarding his personal sentiments toward the scheme's broader urban impact.

HUDSON RIVER BRIDGE AT WEST 125TH STREET

Connecting Cliffside Park, New Jersey, and West 125th Street in Manhattan

Othmar H. Ammann, Chief Engineer
Ammann & Whitney, Consulting Engineers, Engineers of record
Proposal date: 1954
Designed for the Port Authority of New York and New Jersey

In 1954, an ambitious study examining important traffic links in the greater New York area concluded with a series of recommendations to improve traffic flow: adding a lower deck to the George Washington Bridge, building a second bridge between Queens and the Bronx (the Throgs Neck Bridge), and constructing a bridge between Staten Island and Brooklyn (the Verrazano-Narrows Bridge). The study also recommended a second bridge across the Hudson River that would connect West 125th Street in Manhattan to Cliffside Park, New Jersey. Of the above mentioned structures, all were built except the 125th Street Bridge.

Ammann's proposal for the 125th Street Bridge was a two-decked steel suspension structure supported with two pairs of suspension cables resting on rigid-frame towers of steel plates, resembling the Verrazano-Narrows Bridge

A.9 Rendering of the proposed Hudson River Bridge at West 125th Street.
(Rendered by M. Michele; courtesy of Technorama Switzerland)

in form. The span between the proposed bridge's towers was four thousand feet (five hundred feet longer than the George Washington Bridge); distance from anchorage to anchorage was sixty-three hundred feet.

The proposal was not developed beyond schematic design. To this date, no serious attempt has been made by municipalities, state government, or the Port Authority of New York and New Jersey to revive the proposal.

THROGS NECK BRIDGE

Connecting Bayside, Queens, with the Throgs Neck Peninsula in the Bronx

Othmar H. Ammann, Chief Engineer
Ammann & Whitney, Consulting Engineers, Engineers of record
Aymar Embury II, Consulting Architect

Design inception: 1954
Commencement of construction: October 22, 1957
Opening date: January 11, 1961
Owned and operated by the Triborough Bridge and Tunnel Authority

The design of this suspension structure, with its eighteen-hundred-foot span, follows closely that of other bridges of comparable size built at the time—especially the 1957 Walt Whitman Bridge between Philadelphia and Camden, New Jersey, by Modjeski and Masters, Consulting Engineers, for which Ammann and his firm were consulting engineers.

The Throgs Neck's single deck carries six lanes of traffic separated by a four-foot-wide median barrier. The roadway deck consists of a top layer of bituminous concrete resting on five-inch slabs formed by a steel-grid flooring filled with concrete. The slabs in turn rest on continuous longitudinal stringers supported on the top chords of the transverse floor trusses. The transverse trusses are framed into longitudinal stiffening trusses located in the vertical planes of the cables and suspended from the main cables by wire rope. The two lateral members in conjunction with the two vertical stiffening trusses form a rigid frame offering ample resistance against dynamic wind action.

The main cables are twenty-three inches in diameter and composed of 11,400 strands of steel wire. The concrete anchorages at either end measure 140 feet by 200 feet at the base, rising 150 feet above the foundations. The transfer of stress from the cables to the anchorages is made with steel eyebars that, at their lower ends, connect with anchor girders. The girders are about half the length of the eyebars and are embedded in the concrete.

The rigid tower frames are composed of two vertical shafts, a portal strut at the top, and a transverse strut connecting the two legs at the floor level. The tower shafts rest on concrete pedestals that transmit the load to the concrete foundations. The tower legs have a rectangular closed section tapering from twenty feet at the bottom to sixteen feet at the top. The walls and interior partitions of these shafts are formed of solid web plates and angles enclosing a large inside well that houses an elevator. Longitudinally, the tower shafts are flexible enough to bend under the movements of the cable saddles at their tops due to fluctuating temperature and live loads.

Test boring indicated that open-caisson-type foundations for both anchorages and towers would be the most suitable and economical. To reach firm bearing soil, caissons were sunk to 80 and 160 feet, respectively.

SPECIFICATION HIGHLIGHTS
- Total length, including elevated approaches: 13,953 feet
- Suspension span's total length is 2,910 feet (main span is 1,800 feet, and two side spans are 555 feet each)
- Towers rise 355 feet above the mean high-water level
- The center of the bridge's main span clears the mean high-water mark by 142 feet
- The cables contain 13,300 miles of wire

VERRAZANO-NARROWS BRIDGE

Connecting Bay Ridge, Brooklyn, with Fort Wadsworth, Staten Island

Othmar H. Ammann, Chief Engineer
Ammann & Whitney, Consulting Engineers, Engineers of record
George E. Spargo, succeeded in 1963 by Peter J. Reidy, General Project
    Manager for the Triborough Bridge and Tunnel Authority

Design inception: 1948
Commencement of construction: September 1959
Opening date: November 21, 1964
Owned and operated by the Triborough Bridge and Tunnel Authority

The slightly tapered towers of the Verrazano-Narrows Bridge are each built with ten thousand prefabricated steel cells, each of them eight feet wide and forty feet tall. The towers were designed transversely as simple rigid frames, using no diagonal bracing, and relying on lateral strength from the connecting portal struts; the struts are arched at the top of the tower and flat at the connection below the roadway. Foundations for the towers eventually came to rest on firm sand and clay 105 feet below the water surface on the Staten Island side and 170 feet below on the Brooklyn side. The foundations are caissons, poured in successive layers at the water's surface.

The spun galvanized wire of the four main suspension cables—two at each side of the roadway deck—rests on cast-steel saddles near the tops of the two bridge towers. The ends of the cable strands are attached to steel eyebar chains buried in the reinforced concrete of the anchorage.

The suspended two-deck roadway structure is a rigid rectangular tube, with Warren stiffening trusses connecting the decks at the sides. All members of the deck structure act together as a unit to resist vertical, lateral, and torsional forces.

SPECIFICATION HIGHLIGHTS
- Bridge construction used 188,000 tons of steel (three times the amount in the Empire State Building) and 570 cubic yards of concrete
- Weight of steel and concrete (including upper and lower roadways) is 93,500 tons
- Total weight of bridge: 1,265,000 tons, to the Empire State Building's 365,000 tons
- Clear span: 4,260 feet (760 feet longer than that of the George Washington Bridge)
- Side spans: 1,215 feet each
- Length of the suspended structure: 6,690 feet
- Total length of bridge: 13,700 feet
- Roadway soars 228 feet above the Verrazano Narrows
- Pedestals rise 33 feet above water level to support the steel towers
- Each of the two towers rises almost 70 stories (693 feet) above water and required 27,000 tons of steel, enough to make 18,000 automobiles
- The bridge's four main cables are 35 7/8 inches in diameter
- There are over 142,500 miles of galvanized steel wire in the cables—enough to circle the earth approximately six times. The 0.196-inch-diameter wire was unreeled from giant spools—the biggest that had ever

been used in bridge building. Each spool weighed 24 tons and stored 90 miles of wire

· The main cables are each made of 61 strands of 428 wires apiece
· The weight of wire in the four cables is 38,290 tons
· Roadway was constructed from sixty-six 400-ton steel box sections
· Heat and cold expand and contract the cables enough to change the roadway's elevation above the water by 12 feet
· Bridge was designed for 12 lanes of traffic with 237-foot-wide roadways on each deck—annual capacity is 48,000,000 vehicles
· Caisson foundations comprise 155,000 cubic yards of concrete
· Anchorages together contain 780,000 tons of steel and concrete

PONT SUR LA RADE
Geneva, Switzerland

Othmar H. Ammann, Chief Engineer
Ammann & Whitney, Consulting Engineers, Engineers of record
Proposal date: December 1963
Designed for the Swiss Federal Engineering Department

The firm of Ammann & Whitney was hired to assist in the development of a plan for the Geneva Autobahn. As part of the project, Ammann worked in collaboration with the Swiss Federal Engineering Department to prepare a design for a suspension bridge across the harbor basin of Lake Geneva in the city of Geneva. The bridge was critically needed to improve traffic flow and distribution; however, authorization to proceed with the bridge was never given. A bridge or tunnel has still not been built on the proposed site to alleviate Geneva's ever-mounting vehicular traffic congestion.

The proposed Pont sur la Rade is a suspension structure with rigid plate towers. The central span is 2,540 feet—240 feet longer than the Bronx–Whitestone Bridge. Total length of the bridge, from anchorage to anchorage, is 4,720 feet.

Notes

CHAPTER 1: OTHMAR H. AMMANN

Epigraph: Othmar H. Ammann, "A Century of Development in the Building of Steel Bridges," unpublished manuscript for an address given at the Brooklyn Polytechnic Institute to the student chapter of the ASCE, November 18, 1954.

1. Othmar H. Ammann to his parents, Mr. and Mrs. Emanuel Ammann, May 19, 1904; English translation from the original German text by Margot Ammann-Durrer, 1988. In addition to letters written by Othmar and Lilly Ammann, the Othmar H. Ammann Collection housed at the Technorama Schweiz in Winterthur, Switzerland, includes letters written by Mrs. Emanuel Ammann as well as professional correspondence and papers from the estate of Othmar H. Ammann.

2. Ammann to Lilly Selma Wehrli, May 17, 1904; trans. Margot Ammann-Durrer, 1988.

3. Ammann to his parents, August 1906; trans. Margot Ammann-Durrer, 1988.

4. Ammann in diary, March 22, 1923; trans. Margot Ammann-Durrer, 1988.

5. Ammann to his mother, December 14, 1923; trans. Margot Ammann-Durrer, 1988.

6. Quoted in Joseph Gies, *Bridges and Men* (Garden City, N.Y.: Doubleday, 1963), 188.

7. Lindenthal to Silzer, December 20, 1923. Published in Jameson W. Doig, "Politics and the Engineering Mind: O. H. Ammann and the Hidden Story of the George Washington Bridge," *Yearbook of German-American Studies* 25 (1990): 171.

8. Gies, *Bridges and Men,* 243–244.

9. Othmar H. Ammann, "Bridges of New York," *Journal of the Boston Society of Civil Engineers* 32 (1945): 170.

10. Gay Talese, "City Bridge Creator, 85, Keeps Watchful Eye on His Landmarks," *New York Times,* March 26, 1964.

CHAPTER 2: THE GEORGE WASHINGTON BRIDGE

Epigraph: Louis A. Volse, "O. H. Ammann: An Artist in Steel Design," *Engineering News-Record,* May 15, 1958.

1. Le Corbusier, "A Place of Radiant Grace," in *When the Cathedrals Were White* (New York: Reynal & Hitchcock, 1947).

CHAPTER 3: THE BAYONNE BRIDGE

Epigraph: Ammann's address at the opening ceremony of the Bayonne Bridge, published by the Port of New York Authority in a limited and undated edition titled "Bayonne Bridge over the Kill van Kull Between Port Richmond, Staten Island, New York, and Bayonne, New Jersey: Addresses Delivered at Dedication, November 14, 1931."

1. Othmar H. Ammann, "Development of Plans and Research for the Kill van Kull Bridge," text of an address to the annual meeting of the structural division of the American Society of Civil Engineers, January 16, 1930. Condensed in *Proceedings ASCE* 56 (1930): 490.

2. "Bayonne Bridge: Addresses Delivered at Dedication."

3. Ibid.

CHAPTER 4: THE TRIBOROUGH BRIDGE

Epigraph: Othmar H. Ammann, "Bridges of New York," *Journal of the Boston Society of Civil Engineering* 32 (1945): 143–144.

1. Sharon Reier, *The Bridges of New York* (New York: Quadrant Press, 1977), 129.

CHAPTER 5: THE BRONX–WHITESTONE BRIDGE

Epigraph: Othmar H. Ammann, "The Hell Gate Bridge over the East River in New York City," *Transactions ASCE* 82 (1918)

1. Reier, *Bridges of New York,* 136.

2. Ammann, "Present Status of Design of Suspension Bridges."

CHAPTER 6: THE THROGS NECK BRIDGE

Epigraph: Ammann, "A Century of Development in the Building of Steel Bridges."

CHAPTER 7: THE VERRAZANO-NARROWS BRIDGE

Epigraph: Othmar H. Ammann, "Planning and Design of the Verrazano-Narrows Bridge," *Transactions of the New York Academy of Sciences* (1963).

1. Mary Jean Kempner, "The Greatest Bridge of Them All," *Harper's Magazine,* November 1964, 70–76.

2. Gay Talese, *The Bridge* (New York: Harper & Row, 1964), chapter 6.

3. Lewis Mumford, "The New York Skyline," *New Yorker,* November 14, 1959, 186–191.

4. Janice Fioravante, "The Farms Gave Way Just 25 Years Ago," *New York Times,* May 18, 1997.

5. Reier, *Bridges of New York,* 140.

Glossary

abutment upright support designed to resist the weight or force of adjoining structural members. In arch bridges, the abutments take both the weight and thrust of the arch and its roadway.

anchorage a solid mass, usually concrete, into which are fastened the main cables of a suspension bridge

arch bridge a structure that uses a curved beam or curved masonry configuration as its basic load-bearing element

cable bent a structural member used to redirect the inclination of a suspension cable and transfer part of the force carried by the cable

cable saddle a U-shaped seat at the top of a tower of a suspension bridge over which the suspension cables are laid

cable spinning the process of laying strands of wire from anchorage to anchorage until enough wires have been "spun" to be bundled into a suspension bridge's main suspension cable

caisson a watertight box inside which men or machines perform construction work underwater. Caissons were used to excavate riverbeds and then to construct tower foundations for several of Ammann's suspension bridges.

cantilever the portion of a beam or truss that extends beyond a supporting post or wall

cantilever bridge a bridge typically formed by projecting ends of two cantilevers

clear span in suspended bridge design, a segment of roadway under which there is no vertical support

cross bracing diagonal struts used to strengthen and stabilize a structural form

dead load the total weight of a structure and the objects it supports

deadweight synonymous with dead load

deck used herein as an abbreviation for road deck. See *road deck* below.

eyebar a metal strip that is embedded in an anchorage up to an end with an eyelet, through which are passed the wires of a suspension cable during spinning. Eyebars fasten main suspension cables to anchorages.

girder a long beam that can be solid (a plate girder) or latticed

lift bridge a bridge with a movable road deck that slides up and down a vertical bracket

live load dynamic forces that impose internal stresses on a structure that affect its stability (for example, wind or vibration caused by moving vehicles)

piers pillars built beneath bridges for support

portal arch a large-scale gateway formed by an arched entrance

reinforced concrete concrete containing steel bars or mesh to increase its strength

road deck the entire platform assembly that supports a roadway

rocker joint a vertical structural member that connects a road deck to a suspension bridge tower above it, to transfer part of the road deck's dead and live loads to the tower

saddle abbreviation for cable saddle. See *cable saddle* above.

skewback the slanting surface supporting the end of an arch

span the distance between the supports of a bridge

spandrel-braced arch a type of arch structure comprising arched trusses in lateral planes that are tied together with simple trusses in longitudinal planes

spandrel post a vertical structural member that connects the curve of an arch with a horizontal configuration supported from below by the arch

stringer a horizontal structural member that connects upright posts in a truss configuration

strut a brace fitted into a truss configuration to resist pressure in the direction of the strut's length

suspended roadway those portions of a bridge's roadway that are held above ground by an overhead structure

suspender cables wire ropes that connect the road deck to the main suspension cables in a suspension bridge, or to an arch in an arch-type bridge

suspension bridge a bridge suspended from towers, typically having five basic parts: a pair of towers, a road deck, suspension cables, suspender cables, and anchorages. The suspension cable is stretched across the towers and attached to the anchorage. The road deck is hung from the suspension cable by vertical ropes known as suspender cables.

suspension cables primary components of a suspension bridge. Suspension cables are compound bundles of wire stretched from tower to tower and attached to the anchorages by means of eyebars.

suspension chains primary structural members in a suspension bridge that provide an alternative to suspension cables. A suspension chain is composed of long, narrow iron or steel links (usually eyebars) fastened together, draped over the towers of the bridge, and secured in the anchorages.

thrust a pushing force exerted from one object to another

torsion the twisting of an object by holding one end firm and turning the other along a longitudinal axis. Torsion creates internal stresses in an object.

towers the primary vertical structures in a suspension bridge. They carry the suspension cables or chains, transferring most of the deadweight of the road deck, suspender cables, and suspension cables to the ground.

truss a network of "sticks" (known as struts and ties) assembled to form a rigid framework

truss bridge a bridge configuration that relies largely on truss work for its support

viaduct a long bridge with a road, railway, or waterway supported on closely spaced piers or arches

Warren truss a truss with horizontal top and bottom chords connected by diagonals alternately sloped in opposite directions, sometimes modified with the addition of verticals connecting the top and bottom chords. This truss form was invented in England by John Warren and Willoughby Monzani, who were granted a patent on it in 1848.

Selected Bibliography

Ammann, Othmar H. "Additional Stiffening of Bronx–Whitestone Bridge." *Civil Engineering* 16 (1946): 101–103.

———. "Advances in Bridge Construction." *Civil Engineering* 3 (1933): 428–432.

———. "Bridges of New York." *Journal of the Boston Society of Civil Engineers* 32 (1945): 141–170.

———. "Brobdingnagian Bridges." *Technology Review* 33 (1931): 441–444, 464.

———. "A Century of Development in the Building of Steel Bridges." Address given at the Brooklyn Polytechnic Institute to the Student Chapter, ASCE, November 18, 1954.

———. "Development of Plans and Research for the Kill van Kull Bridge." Address to the Annual Meeting of the Structural Division of the ASCE, January 16, 1930. Condensed in *Proceedings ASCE* 56 (1930): 487–493.

———. "The Hell Gate Arch Bridge over the East River in New York City." *Transactions ASCE* 82 (1918).

———. "New Crossings of the Waters of New York." Address given at the Convention of New York State Highway Engineers, New York City, March 1959.

———. "The Planning of Arterial Highways Across the Streams of New York." Address presented at the Stevens Institute, February 10, 1937.

———. "Planning and Design of Bronx–Whitestone Bridge." *Civil Engineering* 9 (1939): 217–220.

———. "Planning the Lincoln Tunnel Under the Hudson." *Civil Engineering* 7 (1937): 387–391.

———. "Possibilities of the Modern Suspension Bridge for Moderate Spans." *Engineering News Record* 90 (1923): 1072–1078.

———. "Present Status of Design of Suspension Bridges with Respect to Dynamic Wind Action." *Journal of the Boston Society of Civil Engineers* 40 (1953): 231–253.

———. "Present Trends in Structural Design." *Civil Engineering* 10 (1940): 21–24.

———. "Verrazano-Narrows Bridge: Conception of Design and Construction Procedure." Address presented at the American Society of Civil Engineers Structural and Engineering Conference and Annual Meeting,
October 19–23, 1964. Reprinted in *Journal of the Construction Division ASCE* (March 1966): 5–21.

Ammann, Othmar H., Collection. Technorama Schweiz. Winterthur, Switzerland. The collection includes professional papers from the Othmar H. Ammann estate and personal letters by him, his first wife, Lilly Selma Ammann (née Wehrli), and his mother, Rosa Ammann (née Labhardt).

Ammann, Othmar H., Allston Dana, Leon S. Moisseiff, et al. "George Washington Bridge Across the Hudson River at New York, N.Y." American Society of Civil Engineers *Transactions* 97 (1933): 1–416. (Discussion of the eight presented papers at 417–442.)

Bowden, E. W. "The Triborough Bridge Project." *Civil Engineering* 6 (1936): 515–519.

Buckley, Tom. "The Eighth Bridge." *New Yorker,* January 14, 1991, 37–59.

Caro, Robert. *The Power Broker: Robert Moses and the Fall of New York.* New York: Vintage, 1975.

Dana, Allston. "Design and Erection—Kill van Kull Bridge." *Proceedings ASCE* 56 (1930): 500–509.

Doig, Jameson W. "Politics and the Engineering Mind: O. H. Ammann and the Hidden Story of the George Washington Bridge." *Yearbook of German-American Studies* 25 (1990): 151–199.

Embury, Aymar, II. "Esthetics of Bridge Anchorages Applying Architectural Principles to Triborough, George Washington, and Whitestone Spans." *Civil Engineering* 8 (1938): 85–89.

Gies, Joseph. *Bridges and Men.* Garden City, N.Y.: Doubleday, 1963.

Kempner, Mary Jean. "The Greatest Bridge of Them All." *Harper's Magazine,* November 1964, 70–76.

McCullough, David. *The Great Bridge.* New York: Simon and Schuster, 1972.

Moisseiff, Leon S. "Design, Material, and Erection of the Kill van Kull (Bayonne) Bridge." *Journal of the Franklin Institute* 213 (1932): 464–502.

"New York's Triborough Bridge." *Engineering News-Record,* August 8, 1935, 177–183.

Petroski, Henry. *Engineers of Dreams.* New York: Alfred A. Knopf, 1995.

Reier, Sharon. *The Bridges of New York.* New York: Quadrant Press, 1977.

Shanor, Rebecca Read. *The City That Never Was.* New York: Viking Penguin, 1988.

Stüssi, Fritz. *Othmar H. Ammann*. Basel: Birkhäuser Verlag, 1974.

Talese, Gay. *The Bridge*. New York: Harper & Row, 1964.

"Throgs Neck Bridge." New York: Triborough Bridge and Tunnel Authority, 1957.

"Verrazano-Narrows Bridge." *Civil Engineering* 34 (1964): 38–50.

Whitney, Charles S. *Bridges: Their Art, Science, and Evolution*. Greenwich, Conn.: Greenwich House, 1983. (First published New York: W. E. Rudge, 1929.)

Widmer, Urs. *Othmar H. Ammann, 1879–1965: 60 Jahre Brückenbau*. Winterthur, Switzerland: Technorama Schweiz, 1979.

Acknowledgments

The individuals and institutions who generously offered support to the project are varied and numerous.

Financial assistance for research, manuscript preparation, and acquiring illustrations was provided by the National Endowment for the Arts' Design Arts Program, the J. M. Kaplan Fund, the Arthur Ross Foundation, the New York Council on the Arts, the Graham Foundation for Advanced Studies in the Arts, Pro Helvetia, Dr. Margot Ammann-Durrer, and Furthermore, the publication program of the J. M. Kaplan Fund.

Sources for the material studied include the New-York Historical Society (my appreciation to Mary Beth Betts and Barbara S. Christen, who cheerfully and patiently assisted the project in the review of hundreds of uncataloged original drawings, photographs, and unpublished written documents); MTA Bridges and Tunnels (my gratitude to Laura Rosen, archivist, whose gracious involvement in the work included research undertaken on the project's behalf); the Port Authority of New York and New Jersey (thanks to Terry Benzic and Connie Delabarka); the Swiss Technorama in Winterthur, Switzerland (Urs Widmer, former president of the Technorama, kindly made the collections accessible and delightfully shared his personal recollections of Ammann); the Prints and Photography Division of the Library of Congress; the New York Public Library, including the Map Division (Alice C. Hudson, chief) and Prints and Photography (Roberta Waddell, curator of prints); the Avery Architecture and Fine Arts Library, Columbia University; and the Museum of the City of New York.

The private collections and personal recollections of several individuals proved invaluable to the study. Dr. Margot Ammann-Durrer, daughter of the engineer, shared personal letters and diaries written in her father's hand, as well as a wealth of anecdotal accounts that together conveyed a deeper sense of a man who kept a low public profile. Edward Cohen and Frank L. Stahl, who for many years were associates of Ammann's, also shared original documents and recounted their memories of the man and the events surrounding the construction of the later bridges.

Ann Foker carefully and patiently assisted with research.

It was necessary to document the bridges photographically. In-kind documentary services were provided by the Historic American Engineering Record (HAER), a division of the National Park Service. HAER assigned an outstanding photographer, Jet Lowe, to work with the project. The Port Authority of New York and New Jersey assisted him with helicopter services. The project is, therefore, much indebted to the Port Authority for assistance with documentation.

I am indebted to scores of individuals who provided knowledge, insight, and moral support during the preparation of this book. Thank you all from the bottom of my heart. I am especially grateful to Rebecca Read Shanor, Joan Pierpoline, Nina Felshin, Michael Dickman, Lora Redford, Katherine Colbert, John Voelcker, Barbara Millstein, Kimberly Blanchard, Dr. Chris Hartman Leubkeman (who extended in-kind assistance through research work conducted in Switzerland), Dianne H. Pilgrim, Susan Yelovich, Diane Littmann, Suzanne Georgy, Eric Delony, Robert Peck, Dave Frieder, Carlos Gibson, Mildred F. Schmertz, and Myron Goldsmith. Dr. Margot Ammann-Durrer most certainly belongs on this list, too: apart from sharing her archive and providing direct financial support, she exercised a sixth sense for knowing when to offer words of encouragement.

Finally, a special note of thanks to the skillful individuals who brought this project to a conclusion: Jane Lebowitz of Mildred Marmur Associates Ltd, and my editors at Yale University Press, Judy Metro, senior editor, and Phillip King, manuscript editor.

Index

(Page numbers in *italic* type refer to illustrations)

abutments, 85, 88, *88*, 92, 167

aerodynamics, 33–34, 124

aesthetics, 35–36, 43, 135; arches and, 82; cable anchorages and, 57, *58*, 59, *59*

airports, 79, 117

air temperature, 138

American Society of Civil Engineers, 11

Ammann, George Andrew (son), 7

Ammann, Klary (second wife, née Vogt), 30, 31

Ammann, Lilly (first wife, née Wehrli), 2, 4, 5, 7, 9, 29

Ammann, Margaret (Margot) (daughter), 30

Ammann, Othmar H., *x*, 1, *8*, 35–38, *38*, 82, 125; and aesthetics, 82, 88, 135; and Bayonne Bridge, 85, 88–92, *165*; and Bronx–Whitestone Bridge, 115, 124; builds career, 4–7; in California, 29–30; catalog of works, 163–175; in civil service, 27–30; and deflection theory, 34, 35; early life in New York City, 1–4; and George Washington Bridge, 43, 45, 49, 57, 59, 63, 163; and Hudson River crossing, 14–16, *16*, 22, 23; Lindenthal and, 7–11, 13, 14–15, 26; marriages, 4, 30; and naval fortifications, 128; patronage and, 25–27; as Port Authority engineer, 26–27, 28, 29, 31, 34–35, 37, 163; and Tacoma Narrows Bridge, 32–34; and Throgs Neck Bridge, 128; and Triborough Bridge, 93, 95, 105, 106, 110–111; on vehicular traffic, 93; and Verrazano-Narrows Bridge, 135, 138, 141, 151

Ammann, Werner (son), 10, 30, 34

Ammann & Combs, Consulting Engineers, 31, 34

Ammann & Whitney, Consulting Engineers, 34–35, 175

anchorages, 57, 93, 106, *107*, 109; of

Bronx–Whitestone Bridge, 169; of George Washington Bridge, *73*, 163–164; of Throgs Neck Bridge, 174; of Triborough Bridge, 166; of Verrazano-Narrows Bridge, 137, 138, 147, *148–150*, 151

approach roads, *118*, 169

arched trusses, *90*

arches, 21, 82, 164, 165; of Bayonne Bridge, 82, *83*, 85

architects, 41, 43

Art Deco, 45, *47*, 169

Arthur Kill, 77

Ash-Howard-Needles and Tammen, Consulting Engineers, 28, 93, 166

Baltimore & Ohio Railroad, 135

baroque design style, 45, *47*

Battery Park, 172

Bayonne Bridge, 27, 37, 77, *78*, *79*, 79–80, *89–91*; compared with Triborough Bridge, 102; form of, 80–82, *81*; masonry abutments, 88, *88*; statistics concerning, 164–165; steel arches and suspended decks of, 82, *83–84*, 85; viaduct piers of, *86*, *87*, 87–88, 92

Bay Ridge (New York neighborhood), 138, 157, 160

Bayside (New York neighborhood), 127

Belt Parkway, 116

Benjamin Franklin Bridge, 135

"boomers," 151

Bradfield, John, 82

bridges: carrying capacity of, 8, 41, 82; civic amenities and, 76; construction of, 106, *106*, 111, *111–112*, 113, *129*, 133, 138, *139*, *149*, 151, *152–156*, 156–157; design of, 4; failure of, 17, *32*, 32–34, 37–38, 120, 170; public plazas and, 63, *64–66*, 66–67, *67–70*, 71; span (length) of, 16, 18, 20, 21, 31, 39, 62, 115, 120, 128, 163,

173, 175; traffic on, 15, 16, 20, 63, 66, *66*

Bronx, 102, 105, 116, 127, *127*, 166

Bronx Kill, 93, 166

Bronx Kill Railroad Bridge, 7

Bronx–Whitestone Bridge, 35, 37, *114*, 115–117, *116–119*, *131*; construction of, 97, 133; design of, 28, 34; span of, 128, 175; statistics concerning, 169–170; structural modifications, 120, *121–123*, 124; vehicular traffic on, 35, 97, 125

Brooklyn, 79, 135, 157, 160, 172, 173

Brooklyn–Battery Bridge, *172*, 172–173

Brooklyn Bridge, 20–21, *22*, 31, 49, 110, 144, 147

Brooklyn–Queens Expressway, 137, *162*

Buck, Leffert L., 21

Burnside Bridge, 13

Byrne, Edward A., 93

cable bents, 109, 147

cables: anchorage of, 17, 57, *58*, 59, *59*; on Brooklyn Bridge, 20–21; cable sag, 34; spun-cable suspension, 49–50, *50–55*, 55, *56*, 57; wire for, 138, 141, *149*. See also anchorages

cable saddles, *107*, *108*, 113, 147, *155–156*, 166

cable spinning, 156

cable stays, 21

caissons, 20, 141, 144, *145–146*, 146, 174

Canada, 5, 6

Canimar River Bridge (Cuba), *170*, 170–171

cantilever bridges, 5, 13, 80, 82

Castro, Fidel, 171

catwalks. See temporary walkways

Central Park (New York City), 71

chain suspension, 18, 23, 49

Chicago, 151

Cincinnati Suspension Bridge, 20

City Beautiful movement, 63

classical design style, 45

clear span, 18, 19, 31, 135; of Bronx–Whitestone Bridge, 115; of Brooklyn Bridge, 20, 21; of George Washington Bridge, 163; roadway width and, 120; of Throgs Neck Bridge, 128; of Verrazano-Narrows Bridge, 138; of Williamsburg Bridge, 21

Clove Lakes Expressway, 160

Columbia River, 4

commuter rail transit, 80

commuters, 80, 97

compression members, 6

Connecticut, 97, 116

Connecticut Society of Engineers, 15

Cross Island Parkway, 116

Dana, Allston, 163

deadweight, 21, 45

decks. See road decks

deflection theory, 18, 21–22, 35; and bridge failure, 33–34; suspension bridges and, 59, 61–63

Delaware Memorial Suspension Bridge, 31, 163

Design of Steel Bridges (Kunz and Schneider), 6

design trends, 45, *46–48*

Eastern Boulevard, 116

East River, 21, 27, 116, 127

East River Drive, 97

Ecole des Ponts et Chaussées (Paris), 4

Ecole Polytechnique, 17

Eisenhower, Dwight D., 125

Eldridge, Clark H., 32

Ellet, Charles, 17–19, 23, 50, 55

Embury, Aymar, II, 59, 105, 109

engineers, 1, 5, 17, 19

eyebars, 19, *44*, 49, *107*, 109, 147, *149*, *150*

ferries, 77, 97, 135
Finley, James, 16–17
Fort Hamilton (Brooklyn), 128, 137
Fort Schuyler (Bronx), 127–128
Fort Wadsworth (Staten Island), 128, 135, 137
Fort Washington (New York neighborhood), 66, 128, 163
Fort Washington Park, 71, *72*
Francis Lewis Park, *119*

"Galloping Gertie," 120
George Washington Bridge, *24*, 37, *40*, 41, *41–42, 74–75, 94*; Ammann's proposal for, 15–16, *16*; cable ends, 57, *58*, 59, *59*; cable spinning, 49–50, *50–55*, 55, *56*, 57; compared with Triborough Bridge, 102; construction of, 39; design of, 1, 27, 28; geography and foundations for, 128; iron workers and, 151; New Jersey Plaza, 67, *67–70*, 71; New York Plaza, 63, *64–66*, 66; parkway approach, 71, *72–73*, 75; road deck of, 59, *60–61*, 61–63, *62*; roadway width ratio to clear-span length, 120; statistics concerning, 163–164; towers, 43, *44*, 45, *45, 46–48*, 49, 92, 110; vehicular traffic on, 35, 39, 41, *42*, 173
Germany, 6, 9
Gilbert, Cass, 45, 49, 63, 71; and ornamentation, 43, 57, 92, 164
girders, 34, 166, 169, 171
girder trusses, 63
Goethals Bridge, 27, 77, 80
Golden Gate Bridge, 33, 36, 138; Ammann and, 29, 30, 163; iron workers and, 151
Gothic design style, 45
Governors Island, 172
Great Depression, 28, 43, 95, 97, 106, 137
guy wires, 59

Hackensack River, 77
Harlem River, 97
Harlem River Pedestrian Bridge, 31, 37, *101*, 163, *171*, 171–172
Hell Gate Arch Bridge, *8*, 12, 27, *94, 96, 99,*

167; Ammann and, 7–9, *8*, 11, 30, 34; monumentality of, 88; trussed arches of, 82. See also Little Hell Gate Railroad Bridge
hemispheric struts, 128
Henry Hudson Bridge, *94*
Henry Hudson Parkway, 71
highways, 93, 117, 127; Bayonne Bridge and, 77, 79, 80; Lincoln Tunnel and, 168; Verrazano-Narrows Bridge and, 135, 160, 162
Hilgard, Karl Emil, 2
Holland Tunnel, 80, 168
Hudson River, 21, 163, 169
Hudson River Bridge (125th Street), 173, *173*
Hudson River bridge project, 11–14, *12, 14*–16, *16*. See also George Washington Bridge
Hutchinson River Parkway, 116
Hylan, John F., 135

Iannnielli, Edward, 157
International school, 41
interstate highways, 80
iron workers, 151, *152–156*, 156–157

John A. Roebling & Sons, 49
Just Such Clay Company, 10, 25

Kill van Kull, 37, 77, 80, 82, 85
Kill van Kull Bridge. *See* Bayonne Bridge
Kunz, Frederic C., 5–7, 29
Kunz & Schneider, Consulting Engineers, 5, 6

"Laboring Group" cable ends, 57, *59*
La Guardia, Fiorello H., 136
Larson, Morgan F., 92
Le Corbusier, 41
"Liberty Bridge," 135
Lincoln Tunnel, 27, 71, *168*, 168–169
Lindenthal, Gustav, 7–11, *8*, 29; Ammann's break with, 14–15, 26; and chain suspension, 23; Hell Gate Bridge project, 34; Hudson River bridge pro-

ject, 11–14, *12;* patronage and, 25; and steel arch design, 82
"Little Green Bridge." *See* Harlem River Pedestrian Bridge
Little Hell Gate Railroad Bridge, 7, 163, *167*, 167–168, 171. *See also* Hell Gate Arch Bridge
Long Island, 41, 77, 97, 116, 168
Long Island Sound, 127
long-span bridges, 37, 39, 170–171; cable spinning for, 50; deflection theory and, 61; design of, 14, 34–35, 124; safety of, 120; towers of, 43; vehicular traffic on, 125

McClintic/Marshall company, 4
McKee, Gerard, 156–157
McKim, Mead, and White, 12
Mackinac Bridge (Michigan), 151, 163
Macombs Dam Bridge, *94*
Madison Avenue Bridge, *94*
Manhattan, 97, 102, 105, 116, 128, 166, 172
Manhattan Bridge, 21, *23*, 61–62
masonry, 43, 88, *88*, 92
Mayer, Joseph, 2, 3, 4, 5, 29
Melan, Joseph, 21
Menai Strait Bridge (Wales), 17, 18, *18*
Merrimack River Bridge (Massachusetts), 17
Miles, Walter, 32
modernism, 45
Modjeski, Ralph, 4, 5, 29, 135
Modjeski and Masters, Consulting Engineers, 173
Moisseiff, Leon S., 115, 120; and deflection theory, 21, 33, 61–62, 124; and trusses, 23, 63
Moses, Robert, 36, 95, 115; and Triborough Bridge and Tunnel Authority, 27–29, 124, 137, 162, 172
MTA Bridges and Tunnels, 97
Mumford, Lewis, 157, 160
Murphy, John "Hard Nose," 151

naval fortifications, 128
neighborhood preservation, 157, 160, *161*, 162, *162*

Newark Bay, 77
New England, 41, 97
New Jersey, 25, 26, 27, 39, 163, 168; Bayonne Bridge and, 77, 79, *79*, 85; George Washington Bridge and, 67, *67–70*, 71; Palisades, 164, 168; Verrazano-Narrows Bridge and, 135
New River Gorge Bridge (West Virginia), 82, 165
New York, upstate, 116
New York City: Ammann's impressions of, 2; Board of Estimate, 137; Board of Transportation, 136; Department of Plants and Structures, 93, 95; Department of Transportation, 167; Fort Washington neighborhood, 66; Parks Department, 168; Pennsylvania Station, 12; World's Fair (1939–1940), 28, 115. *See also* Port Authority of New York and New Jersey
New York Harbor, 37, 77, 135
New York Marathon, *136*
New York State Maritime College, 127–128
Niagara Railroad Suspension Bridge, 20, *20*, 21, 50, 55, 57, 59
Noetzli, Fredi, 30
North River Bridge Company, 11, 12, 13, 14, 25, 26

Olmsted, Frederick Law, 71
145th Street Bridge, *94*
Oregon Trunk Railroad, 4
ornamented abutments, 88, *88,* 92
Outerbridge Crossing Bridge, 27, 77
overhead crossings, 63

Palisades (New Jersey), 82
Passaic River, 77
patronage, bridge engineering and, 25–27
pedestrian traffic, 16, 39, *42*, 102, 106
pedestrian walkways, 15, 167; on Bayonne Bridge, 85, *91*; on George Washington Bridge, *55, 56*; on Triborough Bridge, *100*
Pennsylvania, 16

Pennsylvania Railroad Company, 7, 11, 12
Pennsylvania Station (New York City), 12
Pennsylvania Steel Company, 3, 4, 5
perimeter barricades, 133
Perry, Arthur I., 93, 106, 110, 111
Philadelphia, 6, 9, 16, 17, 151, 173
Philadelphia–Camden Bridge (Benjamin Franklin Bridge), 135
piers, *87*, 87–88
plate girders, 63
plazas, 105, 168, *168*, 169
Polytechnic Institute (Berlin), 19
Pont sur la Rade (Geneva), 175
Port Authority of New York and New Jersey: Ammann as engineer for, 26–27, 28, 29, 31, 34–35, 37, 163; Ammann's retirement from, 171; and Bayonne Bridge, 77, 79, 80, 92; and Throgs Neck Bridge, 125; and unbuilt projects, 173; and Verrazano-Narrows Bridge, 137, 138
Public Works Administration, 95
"punks," 151
"pushers," 151

Quebec Bridge, 5, 6
Queens, Borough of, 102, *104*, 105, 116, 127, 166
Queensboro Bridge, 5, 97, 106
Queens–Midtown Tunnel, 168

railroad bridges, 7, 10
railroads, 13–14, 20, 80, 116
ramps, 63, 66
Randall's Island, 93, 95, *95, 96*, 102, 165, 166, 167; interchange plaza on, 102, *103*, 105
Rea, Samuel, 12
regional development, 157, 160, *161*, 162, *162*
Regional Plan Association, 28, 116
Reidy, Peter J., 151
reinforced concrete, 87
residential districts, 116
Riverside Drive, 71, *72*
Riverside Park, 71, *72*
road decks, 16, 17, 18, 28, 124; of Bayonne

Bridge, *84*, 85, 165; of Bronx–Whitestone Bridge, *117*, 120, 170; deflection theory and, 33–34, 59, 61–63; of George Washington Bridge, *60*, 164; of Throgs Neck Bridge, *129–130*, 173; of Triborough Bridge, *101, 105*, 106, *106*, 166; of Verrazano-Narrows Bridge, 35, *142–143*, 174
Rockefeller, Nelson A., 138
Roebling, John A., 19–20, 23, 59
Roebling, Washington, 20, 144
Roosevelt, Franklin D., 172
Ross Island Bridge, 13
Rowland Prize, 11

saddles. *See* cable saddles
St. John River, 6
St. Lawrence River, 5
San Francisco, 30
Schneider, Charles C., 5
Schuylkill River Bridge, 17–18, *18*
Scientific American, 12, 13
Sciotoville Bridge, 10
Seventh Avenue Bridge (Pittsburgh), 16
Ship Canal Bridge, *94*
shipping lanes, 85, 87
Silzer, George S., 25–27
skyscrapers, 2, 12, 43, 45, 151
spandrel braces, 82, 164
Spargo, George E., 151
stabilization, 17
Staten Island: Bayonne Bridge and, 77, 79, 80, 82, 85; Fort Wadsworth, 128; Verrazano-Narrows Bridge and, 135, 138, 160, 173, 174
Staten Island Expressway, 137
steel, 1, 2, 3, 62; plate girders, 102, 115; towers, 21, 43, 45, *48*, 164; World War II and, 171
Steinman, David B., *8*, 9, 135, 138
Strauss, Joseph B., 163
stress, physics of, 17
struts, 174
suspension bridges, 16–21, *18–20, 22–24*, 80; anchorages of, 147; road decks of, 59, 61–63; Triborough Bridge, 105–106, *106, 107–109*, 109; Verrazano-

Narrows Bridge, 35, 138; Warren trusses and, 82
suspension cables: on Bronx–Whitestone Bridge, 124; on George Washington Bridge, 164; on Hudson River Bridge (125th Street), 173; on Throgs Neck Bridge, 133; on Triborough Bridge, 106, *107*, 109; on Verrazano-Narrows Bridge, 147
Switzerland, 2, 6, 10, 30; Ammann's family in, 7, 9, 14, 15; Federal Polytechnic Institute, 1, 30; Pont sur la Rade (Geneva), 175
Sydney Harbor Bridge (Australia), 82, 165

Tacoma Narrows Bridge, *32*, 32–34, 120, 124, 170
Talese, Gay, 38
Telford, Thomas, 17
temporary falsework construction method, 85
temporary walkways, 49, *50, 51, 155*
Third Avenue Bridge, *94*
Throgs Neck Bridge, 37, 125, *126–127*, 127–128, *131–132*; construction of, 35, 97; statistics concerning, 173–174; strength of, 128, *129*, 133
toll plazas, *69, 70*, 71, *158–159*
tolls, 95, 105
torsion, 133, 141, 164
towers, 20, 28, 35, 128; all-steel, 21; of Bronx–Whitestone Bridge, 169, 170; of George Washington Bridge, *40*, 43, *44*, 45, *45–48*, 49, 76, 164; of Throgs Neck Bridge, 174; of Triborough Bridge, 93, 106, *109, 110*, 110–111, *111–112*, 113, 166; of Verrazano-Narrows Bridge, 137, 138, *140*, 141, 144, *154, 161*, 174
traffic, 20, 63, 173; on Bayonne Bridge, 80; destabilizing effects of, 124; on George Washington Bridge, 39, *42*; increases in, 125, 127; Lincoln Tunnel and, 168; on Throgs Neck Bridge, 133; on Triborough Bridge, 97, 106; on Verrazano-Narrows Bridge, 175. *See also* vehicular traffic

transferral of force, 147
Triborough Bridge, 37, 41, 93, *94*, 95, *96*, 97; construction of, 27–28, 133; lift bridge, 97, *98–101*, 102; statistics concerning, 165–166; suspension bridge, 59, 105–106, *106, 107–109*, 109; towers, *110*, 110–111, *111–112*, 113; truss bridge, 93, 102; viaduct bridge and interchange, 102, *103–104*, 105
Triborough Bridge and Tunnel Authority, 31, 35, 37, 125, 127, 167; Ammann's retirement from, 171; Robert Moses and, 27, 29, 124, 137, 162, 172; as Triborough Bridge Authority, 95, 172
truss bridges, 13, 27
trusses, 16, 18, 21, 34, 97, 102; of Bayonne Bridge, 85; and stabilization of road decks, 59, *60, 61*; on Throgs Neck Bridge, 133; on Verrazano-Narrows Bridge, 141
tunnels, 37, 79
Twain, Mark, 20

underpasses, 63, *73*
University Heights Bridge, *94*

vehicular traffic, 15, 16, 93, 175; on Bayonne Bridge, 80, 87; on George Washington Bridge, 39, *42*, 63, 66, *66*; on Triborough Bridge, *100, 101*. *See also* traffic
Verrazano-Narrows Bridge, 35–36, *36*, 37, 38, 79, *134*, 135–138, 173; as Ammann's final work, 138, 141; anchorages, 59, 147, *148–150*, 151; caisson foundations, 141, 144, *145–146*, 146; construction of, 133, 151, *152–156*, 156–157; geography and foundations for, 128; neighborhood preservation issues and, 157, 160, *161, 162, 162*; New York Marathon and, *136*; road decks, *142–143*; statistics concerning, 35, 174–175; toll plazas, *158–159*
viaducts: of Bayonne Bridge, 80, *81, 86, 87*, 87–88, 165; of Bronx–Whitestone Bridge, 169, 170; of Triborough Bridge, 27, 95, 102, *103–104*, 105, 166

Vogt, Klary. *See* Ammann, Klary
Von Karman, Theodore, 33

Waddell, J. A. L., 27
Waddell and Hardesty, Consulting Engineers, 27, 77
"walkin' bosses," 151
walkways, 106, 167
Walsh, Homer, 50
Walt Whitman Bridge, 151, 163, 173

Wards Island, 93, *95, 96,* 102, 166, 167, 172
Warren trusses: on Bronx–Whitestone Bridge, 115, 120, 124, 170; suspension bridges and, 82; on Triborough Bridge, 106, 166; on Verrazano-Narrows Bridge, 174
Washington Bridge, *94*
Wehrli, Lilly Selma. *See* Ammann, Lilly
Westchester County, 41, 49, 97, 116

West Side Highway. *See* Henry Hudson Parkway
Wheeling Suspension Bridge (West Virginia), 18, *19*
Whitney, Charles S., 34
Willamette River (Oregon), 13
Williamsburg Bridge, 21
Willis Avenue Bridge, *94, 101*
Wilson, Woodrow, 25
wind: and bridge failure, 17, 32, *32,* 33–34,

120; deflection theory and, 21, 59, 63; stiffening trusses and, 133, 141, 173
wind tunnel testing, 33, 124
Winged Tire cable ends, 57, *58*
wire-cable suspension, 18, 50, *50,* 55
Woodruff, Glenn B., 33
Works Progress Administration (WPA), 29, 169
World War I, 9, 10, 11, 13
World War II, 34, 171